Aleksandr Porfir'evich Borodin
Portrait by E[lena] T[imofeevna] Makovskaia (1870)
(The original is to be found in the Glinka State Central Museum of Musical Culture, Moscow. — T.)

N. A. Figurovskiĭ and Yu. I. Solov'ev

Aleksandr Porfir'evich Borodin

A Chemist's Biography

Translated from the Russian
by Charlene Steinberg and George B. Kauffman

Foreword by Martin D. Kamen

Springer-Verlag Berlin Heidelberg New York
London Paris Tokyo

Translators:

Charlene Steinberg
Department of Chemistry
University of Wisconsin Center-Sheboygan
Sheboygan, WI 53081, USA

George B. Kauffman
Department of Chemistry
School of Natural Sciences
California State University Fresno
Fresno, CA 93740, USA

Author of Foreword:

Martin D. Kamen
Professor Emeritus of Chemistry
University of California, San Diego
La Jolla, CA 92093, USA

ISBN-13:978-3-642-72734-4 e-ISBN-13:978-3-642-72732-0
DOI: 10.1007/978-3-642-72732-0

Library of Congress Cataloging in Publication Data.
Figurovskiĭ, Nikolaĭ Aleksandrovich. Aleksandr Porfir'evich Borodin.
Translation of: Aleksandr Porfir'evich Borodin.
Bibliography: p.
Includes index.
1. Borodin, Aleksandr Porfir'evich, 1833–1887. 2. Chemists—Soviet Union—Biography.
I. Solov'ev, I͡Uriĭ Ivanovich. II. Title.
QD22.B72F513 1988 540'.92'4 [B] 87-23433 ISBN-13:978-3-642-72734-4 (U.S.)

Softcover reprint of the hardcover 1st edition 1988

2152/3020-543210

This book is dedicated to the cause
of friendship between two great peoples,
those of the
United States of America
and those of the
Union of Soviet Socialist Republics

Aleksandr Porfir'evich Borodin

by N. A. Figurovskiĭ and Yu. I. Solov'ev

Excerpts from reviews of the Russian Edition (1950)

"... completely reveals the charming aspect of a brilliant representative of Russian culture in a lively, popular form ...

... provides great interest for broad circles of Soviet readers ...

... Borodin's letters to chemists and letters from other scientists received by him, which are [here] published for the first time, are of great interest. Borodin's complete letter to A. M. Butlerov is published for the first time. The author [Borodin] writes with bitterness about the difficulties in building a new laboratory at the Military-Medical Academy and about the absurd organization of work, which hampered fruitful scientific and pedagogical activity ...

... Borodin's life is depicted against a background of the cultural life of Russia."

— O. A. Starosel'skaia-Nikitina: Sovetskaia Kniga. *3*: 81 (1951)

"... fully elucidates for the first time the scientific activity of [this] outstanding Russian chemist, establishing the contemporary significance of his works ... a valuable essay on the history of Russian science written in a popular and fascinating manner. The authors make the most of the works of Borodin and his students, the large collection of archival materials, and the epistolary legacy of the last century. It is the first monograph which acquaints the reader with all sides of the life and versatile activities of [this] outstanding Russian scholar-patriot and great composer."

— V. V. Razumovskiĭ: Priroda. *3*: 87, 88 (1951)

Foreword

A full century has passed since the sudden and tragically premature demise of Aleksandr Porfir'evich Borodin in 1887 at the age of 53, when he was following with phenomenal success the disparate careers of musician, composer, organic chemist, and pioneer in women's medical education. As a unique figure among the remarkable group of geniuses who suddenly appeared in Russia in the middle of the last century and explosively propelled that country into the mainstream of world culture in the arts, humanities, and sciences, it might have been expected that Borodin was the object of much research. However, he remains a victim of an affliction on-going in the biographical arts – the Missing Image Syndrome, as I have had the presumption to term it. In the production of a color photograph the conventional procedure requires the preparation of three images as seen through three primary colors. These are superimposed to give a photograph or projection which shows the original object faithfully rendered in full color. The loss of any one of these partial color images results in a distorted color rendition. The analogy to the situation confronting the reader supplied with an essentially one-sided biography is clear. For example, there are few who would be satisfied with a biography of Leonardo da Vinci written wholly from the aspect of his interest and achievements in aeronautical science.

In the Western world – and even in Russia – the predominant documentation of Borodin's activities has been concerned with his musical genius. There is no doubt that he was one of the supremely talented melodists in musical history, and a testimonial to this gift is readily at hand in the popularity of the many exotic themes that he bequeathed in his compositions, adapted by Robert Wright and George Forrest for their musical comedy *Kismet*. But, in accord with the syndrome mentioned above, his images as a chemist and educator are missing. That of the chemist was rescued from oblivion only recently in Russia, with the appearance of the first and only book-length biography on record dealing mainly with Borodin as a great pioneer of Russian organic chemistry. However, as its authors – Professors N. A. Figurovskiĭ and Yu. I. Solov'ev – explicitly state, this memoir was projected as one in a series of short monographs intended for popular consumption. Still, it went some distance in closing the yawning gaps in the biographical treatment of Borodin's life and accomplishments.

There is no doubt that the Russian contribution to the amazing development of structural chemistry in the last century has tended to be underplayed, while that in the rest of Europe has received much more attention. One wonders, in particular, whether Borodin's name might not have appeared in the chemical pantheon, as have those of Mendeleev, Markovnikov, Menshutkin, and many other

Russians, if the aldol condensation, which he was the first to discover and investigate, had been named the Borodin condensation. Straightening out the record is important; Figurovskiĭ and Solov'ev's biography does much in this respect. Just as meritorious have been the scholarly and exhaustive efforts of Professors Charlene Steinberg and George B. Kauffman, who have made the Russian text accessible to the Western world in their accurate and engrossing translation. Their task was complicated not only by the well-known difficulties in producing an English text completely faithful in spirit as well as detail to the Russian version but also by the somewhat strident propagandist overtones which to some extent have weakened the generally favorable impression of the case that the Russian authors were making to uphold the priority of Borodin's researches in various areas of aldehyde chemistry. The final version now offered to the reader seems eminently adapted to permit accurate evaluation of the merits of the Russian text.

What emerges is a document which for the first time enables the chemist and also the intelligent lay reader outside Russia to appreciate the remarkable contributions to organic chemistry as well as to women's medical education made by Borodin, despite infuriating distractions, his natural inability to say "no", and by pathetic inadequacies in facilities and instrumentation. What can also be seen is the scope of the many questions remaining to be answered. It is clear that a treatment in depth of the Borodin corpus of chemical research is still to be done. For the psychohistorian there is the need to probe Borodin's complex interactions with his friends and particularly with his remarkable mother and neurasthenic wife. One receives a strong impression that Borodin accomplished as much as he did, despite his natural indolence, because of the dedication of his mother, who kept him focused on achievement rather than on nondirectional activity.

But this is not the occasion to indulge in speculations as to what professionals in the area of cultural history and biography will find when they start mining the resource materials which hopefully still exist. One can only wish that the present translation will arouse attention among a wider audience than its original Russian version so that eventually there will be available biographies which show Borodin in all his images and in full color.

Montecito, May 1988 Martin D. Kamen

Dr. Kamen, codiscoverer of carbon-14, the critical long-lived isotope that revolutionized the medical and biological sciences through its use as a tracer, was a central figure in the pioneering nuclear research performed at the Radiation Laboratory of the University of California, Berkeley. He attained international recognition for his work on bacterial photosynthesis, bacterial cytochromes, and the use of radioisotopic tracers in biological and biomedical research. He was born in Toronto to Russian immigrant parents; like Borodin, he is an accomplished musician, who plays chamber music and symphonic music. His latest book is *Radiant Science, Dark Politics: A Memoir of the Nuclear Age* (University of California Press, 1985).

Translators' Preface

The centenary of Borodin's death [he died on February 15 (Old Style), February 27 (New Style), 1887] provided the impetus for our translation of the Russian monograph *Aleksandr Porfir'evich Borodin*,[1] written as part of the Academy of Sciences of the USSR's Nauchno-populiarnaia seriia (Popular Scientific Series) by two distinguished Russian historians of chemistry, Professors Nikolaĭ Aleksandrovich Figurovskiĭ and Yuriĭ Ivanovich Solov'ev of the Academy's Institute for the History of Science and Technology. This 212-page book, published 38 years ago but hitherto inaccessible to chemists, historians of science, and music lovers who do not read Russian, filled a gap in the literature by dealing with both Borodin's chemical and scientific work as a whole in the milieu of nineteenth-century Russian life. Since English has become a virtual lingua franca in the course of the last few decades, we hope that the fruit of our effort will provide a source of information on Borodin for an international audience.

Cognizant of the Italian warning "traduttori traditori" (translators are traitors), we have adhered as closely as possible to the original text, but we have not hesitated to change words here and there, to condense repetitive or circumlocutious expressions, or to recast unusually awkward constructions into idiomatic English. Our goal has been to preserve the spirit rather than the letter of the text, and we hope that the result will approximate a book written by the authors as if English were their native language rather than a translation.

We have also introduced a number of features that should make the volume more useful and convenient for Western readers, e.g., translators' footnotes (designated "T." to avoid confusion with the authors' footnotes, designated "A.") for unfamiliar terms; the use of first names and patronymics the first time that a proper name appears and then usually omission of initials thereafter; names of publishers when available for references cited; an index of Borodin's musical compositions; name and subject indexes; and a selected bibliography. In many cases we have included transliterated Russian terms, followed by English equivalents in parentheses. Material added by the translators is usually enclosed in brackets.

---

[1] Н. А. Фигуровский и Ю. И. Соловъев: Александр Порфиръевич Бородин. Издательство Академии Наук СССР, Москва-Ленинград, 1950;
N. A. Figurovskiĭ i Yu. I. Solov'ev: Aleksandr Porfir'evich Borodin. Izdatel'stvo Akademii Nauk SSSR (Publishing House of the Academy of Sciences of the USSR), Moscow—Leningrad, 1950.

There are numerous conventions concerning the transliteration of Cyrillic characters. We have consistently employed the Library of Congress System II[2] except for proper nouns which have a conventional English spelling, e.g., Czar not Tsar, Beilstein not Beĭl'shteĭn, Goldstein not Gol'dshteĭn, Herzen not Gertsen, Mariinsky not Mariinskiĭ, Mussorgsky not Musorgskiĭ, Prince Igor not Prince Igor', Rimsky-Korsakov not Rimskiĭ-Korsakov, Stassov not Stasov, Tchaikovsky not Chaikovskiĭ, Tolstoy not Tolstoĭ, etc. We have also rendered Ю and Я at the beginning of a word as Yu and Ya, respectively, rather than Iu and Ia.

Since the USSR did not adopt the Gregorian calendar (New Style – N.S.) until February 1, 1918, Russian dates (Julian calendar or Old Style – O.S.) at the time of the events taking place in this book were 12 days behind those in other European countries. In this book the reformed calendar (N.S.) is used for events occurring outside Russia, while the unreformed calendar (O.S.) is used for events inside Russia. In cases of possible ambiguity both forms are given.

We are indebted to Professor Yuriĭ Ivanovich Solov'ev, Institute for the History of Science and Technology, Academy of Sciences of the USSR, Moscow; Professor Ian D. Rae, Department of Chemistry, Monash University, Clayton, Victoria, Australia; Professor Emeritus Martin D. Kamen, Department of Chemistry, University of California, San Diego, La Jolla, California; and Professors Richard P. Ciula and Ronald L. Marhenke, California State University, Fresno; for critical readings of the manuscript. Of course, we are solely responsible for any distortions or errors that we may have unwittingly introduced in the course of our translation. We also wish to thank Sharon Turner and Mary McIntyre for typing the manuscript while simultaneously coping with the idiosyncrasies of new word processors. We are indebted to Robert Michelotti for reproducing the photographs in this volume. Furthermore, we gratefully acknowledge the assistance of the following librarians of the Henry Miller Madden Library of California State University, Fresno – A. Zane Clark, Thomas J. Ebert, Herbert S. Fox, William H. Heinlen, Diane L. Majors, Paul M. Priebe, and Jeanne M. Tempesta.

Last but not least, one of us (GBK) wishes to thank the California State University, Fresno, for a sabbatical leave of absence.

Charlene Steinberg George B. Kauffman

[2] Joseph Thomas Shaw: The Transliteration of Modern Russian for English-Language Publications. University of Wisconsin Press, Madison, WI, 1967, p. 8.

Contents

List of Illustrations

Preface

Aleksandr Porfir'evich Borodin belongs among the outstanding Russian men of science and art of the second half of the nineteenth century. He was a prominent scholar-chemist, a spokesman for the brilliant galaxy of Russian chemists of the 1860–1880's who enriched world science by discoveries of paramount significance. He was at the same time one of the most lucid and popular spokesmen of the "Mighty Little Group,"[1] who, after the genius Mikhail Ivanovich Glinka,[2] laid the true foundation of Russian musical culture.

A vast amount of literature is devoted to Borodin's life and works. However, it is impossible to find any satisfactory interpretation of Borodin's activities as a chemist in this literature. Perhaps Borodin's obituary, written by his closest student, Aleksandr Pavlovich Dianin, professor of the Medical-Surgical Academy (now the S. N. Kirov Military Medical Academy) is the sole exception. In this obituary Borodin's principal researches in organic chemistry are enumerated, and a bibliography of his scientific works is given. In all other publications devoted to Borodin his scientific work is mentioned in only the most general way.

Yet, during Borodin's lifetime great Russian chemists such as Dmitriĭ Ivanovich Mendeleev,[3] Nikolaĭ Nikolaevich Zinin, Aleksandr Mikhaĭlovich Butlerov, and many others ranked him as a first-class scholar to whom chemistry was greatly indebted. Twenty original scientific investigations, carried out by Borodin during the period from 1858 to 1877, won him wide fame in the scholarly world not only in Russia but also abroad during his lifetime. For example, in his memoirs Mendeleev noted that during the time of his journeys abroad, upon meeting with foreign chemists he often heard the question, "Now, what has your Borodin done that's new?" Nevertheless, to a considerable degree Borodin was forgotten as a chemist shortly after his sudden death, thereby sharing the fate of many outstanding Russian scholars of past generations.

A large number of works are devoted to Borodin's musical works in our literature. His inspired productions – his epic opera *Prince Igor*, The Bogatyr (Heroic) Symphony [Symphony No. 2, in B Minor], quartets, and songs – are beloved and popular throughout the broadest sections of our population.

One of Russia's outstanding musical critics, V. V. Stassov,[4] wrote:

> Borodin wrote little from a quantitative point of view,[5] much less than his comrades, but his works, almost without exception, carry the stamp of complete development and profound perfection. There are no weaknesses among them ... Borodin's talent is equally as mighty and striking in a symphony as in an opera or song. His principal qualities are gigantic force and breadth,

tremendous range, dash, and beauty (V. V. Stassov: Biografiia A. P. Borodina (Biography of A. P. Borodin). Istoricheskiĭ Vestnik (Historical Herald), St. Petersburg, [*28*:] 167 (1887).

It is not surprising that Borodin's activity as a chemist should be eclipsed by his work as a composer. His chemical works were relegated to comparative obscurity. However, even during Borodin's lifetime many of his musical friends asserted that his occupation with chemistry interfered with his activity as a composer. They often pointed out to him that the time that he spent in chemical pursuits was a loss for music and that composition rather than research should be his main task in life.

In order to recreate a sufficiently complete view of this outstanding representative of Russian culture it is necessary to explain both sides of his creative activity. Therefore the authors of the present work have conscientiously tried not to separate the description of the chemical activities of this remarkable representative of the 1860's from his musical activities.

In addition to his scientific and musical work, Borodin devoted much time and effort to pedagogical and public activities. Only now is it possible to appraise fully the enormous efforts that Aleksandr Porfir'evich expended in establishing the proper teaching of chemistry in the Medical-Surgical Academy.

At that time there were no good, organized practical courses in chemistry for students in the schools of higher education. Before completing their education only a small number of chosen students had the honor of working in a laboratory under the guidance of a professor. Borodin was the first to organize practical work for 300 to 400 students in the Medical-Surgical Academy. In addition, he availed himself of every opportunity to give the future medical doctors a full measure of theoretical knowledge of chemistry.

Borodin established a new method of teaching by lectures. Combining strictness and accuracy with the use of visual aids, he carried out numerous lecture-demonstrations and enriched the chemical laboratory of the Medical-Surgical Academy with new demonstration devices and preparations.

Borodin devoted himself entirely to his pedagogical activity. He worked with dedication and without rest. In one of his letters to his wife he wrote:

> There is nothing to write about myself – everything is the same: the lectures are now in full swing, the experiments in the laboratory are in great demand; then there are exams, commissions, conferences ... there is no time to breathe.

Borodin's talent as a teacher was particularly apparent in connection with his activity in the women's medical courses. Founded on the initiative and with the spontaneous collaboration of prominent Russian scholars, these courses appeared as the first example of such training in Russia and in the West. The organizers of the courses, among whom Borodin takes no back seat, had to expend much energy to pierce the wall of conservatism of officials and of reactionary higher circles. The Czarist government feared that higher education of Russian women would lead to the development of a revolutionary movement in the country, and it opposed in every possible way the opening of the women's higher medical courses. Only as a result of the incredible growth of epidemic diseases and the terrible child mortality rate did the problem of training women doctors become especially acute, and women's medical courses were first established in 1872.

Together with other prominent Russian scholars, Borodin contributed vigorously to teaching these courses. His many years of activity in this field were very fruitful. The first Russian women doctors always remembered their professor with thanks and love. After Borodin's death the women doctors placed a silver wreath on his grave with the inscription:

> To the founder, protector, champion of the School of Medicine for Women, to the supporter and friend of the students – the women doctors of ten graduations 1872–1878.

Borodin was the successor in the great patriotic tradition of the leading Russian scholars, who spared no effort in training a cadre of well-educated men, a difficult undertaking in Russia at that time. Like the "Grandfather of Russian Chemists", Aleksandr Aleksandrovich Voskresenskiĭ, and many prominent professors of Russian universities – D. I. Mendeleev, A. M. Butlerov, Ivan Mikhaĭlovich Sechenov, and others – Borodin considered his pedagogical activity as a scholar's responsibility no less important than his experimental scientific work.

The efforts of these men of the Russian enlightenment were not wasted. Our country is indebted to these scholars for the rise of numerous schools of talented naturalists, who will forever be an honor to Russian science.

This book is based on a study of all accessible sources of Borodin's life and works. We also became familiar with archival material on Borodin's activities and with his unpublished letters. Sergeĭ Aleksandrovich Dianin and V. A. Kiselov greatly helped by preparing the edition of Vol. IV of Borodin's letters, as did O. P. Lamm, who compiled the detailed bibliography of Borodin's chemical works. Professor I. F. Belza prepared a series of instructions concerning the chapters devoted to music. To all these the authors give their deepest thanks.

Realizing that in a short biographical sketch it is difficult to picture completely the varied creative activities of this remarkable Russian scholar and composer, we have endeavored to depict, if only in the most general manner, the image of one of the leading figures of Russian culture – an outstanding scientist and public figure of the 1860's.

The authors do not claim any complete interpretation of Borodin's activities as a composer, but they do endeavor to deal with his lesser-known works as a chemist, scientist, and public figure.

Footnotes to Preface

[1] Often variously referred to as The Mighty Five, Great Five, The Five, etc. In 1867 the Russian musical critic Vladimir Vasil'evich Stassov (1824–1906) coined the term Moguchaia Kuchka meaning mighty little heap. The group was never referred to as The Five in Russia. For an essay on Stassov, who played an important role in Borodin's career, see Gerald Abraham: V. V. Stasov: Man and Critic in Vladimir Vasilevich Stasov: "Selected Essays on Music," translated by Florence Jonas, Barrie & Rockliff, The Cresset Press, London, 1968, pp. 1–13. — T.

[2] See David Brown: Mikhail Ivanovich Glinka. In: Sidney Sadie (ed.): The New Grove Dictionary of Music and Musicians (hereafter abbreviated as Grove). Macmillan Publishers Ltd., London, 1980, Vol. 7, pp. 434–446; Gerald Abraham: Glinka and His Achievement. In: Gerald Abraham: Studies in Russian Music. William Reeves, London, [1935,] pp. 21–42; Gerald Abraham: Michael Glinka. In: Michel D. Calvocoressi and Gerald Abraham: Masters of Russian Music. Duckworth, London, 1936, pp. 13–64; Boris Vladimirovich Asaf'ev (Igor Glebov): Glinka. Muzgiz, Moscow, 1947; V. V. Stasov: Mikhail Ivanovich Glinka. Gosudarstvennoe muzykal'noe izdatel'stvo, Moscow, 1953; and David Brown: Glinka: A Biographical and Critical Study. Oxford University Press, London, New York, 1974. — T.

[3] For a recent biography see Nikolaĭ A. Figurovskiĭ: Dmitriĭ Ivanovich Mendeleev, 1834–1907, 2nd edition Nauka, Moscow, 1983. — T.

[4] See Gerald Abraham: Vladimir Vasil'yevich Stasov. In: Grove, 1980, Vol. 8, pp. 83–84. — T.

[5] In 1906 the British music educator Sir Henry Hadow wrote, "No musician has ever claimed immortality with so slender an offering – yet if there be, indeed, immortalities in music, Borodin's claim is incontestable." — T.

Introduction

Borodin and His Time

The epoch to which Aleksandr Porfir'evich Borodin belongs was the period in which capitalism developed, breaking the ancient yoke of serfdom and feudal oppression. It was an epoch in which the most profound economic and political opposition, which had already been ripening for a long time, was clearly revealed.

The creative activity of Borodin, one of the greatest Russian composers and at the same time an outstanding chemist, encompasses a 60-year period of the last century. It was a period when, under conditions of revolutionary advancement and under the ideological influence of revolutionary democrats such as Aleksandr Ivanovich Herzen, Vissarion Grigor'evich Belinskiĭ, Nikolaĭ Gavrilovich Chernyshevskiĭ, and Nikolaĭ Aleksandrovich Dobroliubov, there sprang up the Russian scientific schools of militant materialists who were fighting for the advancement of science and for the enlightment of their people.

Profound economic causes lay at the root of this social uprising, which affected all sides of Russian life, science, art, and literature.

Russia started on the road of capitalistic development later than other European countries. Only at the beginning of the nineteenth century did rapid growth of industry begin in Russia. Incompatible opposition arose between capitalism, which had just energed, and the old economic structure of a government which operated under the yoke of serfdom and feudal oppression.

Serfdom in Russia obviously hindered the development of capitalistic elements. V[ladimir] I[l'ich] Lenin wrote:

> ... that form of landed property which one finds in beginning history to develop the capitalistic mode of production *does not correspond* to capitalism. Capitalism itself creates corresponding forms of land relations from old forms, from feudal landowners, from peasant communes, from clan landowners, etc.[1]

If, during the beginning of the nineteenth century, the conflict between the rising capitalism and the serf-feudal system had expressed itself in an individual statement of the revolutionary representatives of Russian society (Decembrists)[2], by the middle of the nineteenth century these conflicts would already have produced a wide social division in the country. In many provinces peasant agitations blazed forth, enveloping all the central provinces of Russia in the 1850's and 1860's. The revolutionary movement sprang up among the prominent Russian intelligentsia.

The defeat of Czarism in the Crimean War[3] (1853–1856) and the rising peasant agitations, which established a threat to the liberation of the peasants "from below", made the refusal of "reforms" impossible for the Czarist government at the beginning of the 1860's.

Lenin clearly characterized the peasant reform of 1861, calling it serf-owner reform:

> Notorious "emancipation" was the most unscrupulous robbery of the peasants; it was a series of violent acts and complete outrages against them . . . In general, the whole "epoch of reforms" of the 60's left the peasants beggars, forgotten, ignorant subordinates to the landowners both at court and in government, both in school and in the Zemstvo[4,5].

However, with all its serf-owner character, the peasant "reforms" of 1861 opened the gates for the growth of capitalism in Russia.

> *As long as* the peasant escaped from the authority of the landlord, *as long as* he remained under the authority of money, he found himself under conditions of commodity production, which proved to be dependent upon the rising capital. And after 1861 the growth of capitalism in Russia proceeded with such speed that in several decades changes took place which required an entire century in several old countries of Europe[6].

In its turn the stormy growth of industry influenced the development of science and technology.

The role of the revolutionary democrats Belinskiĭ, Chernyshevskiĭ, and Dobroliubov as ideological leaders of the prominent Russian intelligentsia was very great. Propagandizing the "idea of peasant revolution, the idea of the struggle of the masses for the overthrow of all old authority"[7] and fighting for public education, the revolutionary democrats raised the banner of struggle against autocracy in political authority and philosophical and scientific authority.

All of the prominent sectors of Russian society responded to Belinskiĭ, Herzen, and Chernyshevskiĭ's fiery calls to study nature. The literature of natural science, boldly raising questions of understanding the material world, enjoyed an unprecedented success among the masses of the prominent revolutionary-minded Russian intelligentsia. Pis'ma ob izuchenii prirody (*Letters on the Study of Nature*) by Herzen, *Origin of Species* by Darwin, articles discussing questions of natural science in popular journals such as Sovremennik (The Contemporary) or Russkoe Slovo (Russian Word), and the artistic works of [Aleksandr Ivanovich] Nekrasov, Chernyshevskiĭ, and other great writers cultivated a revolutionary, materialistic attitude in Russian youth.

The question of the growth of *native* science appeared in articles by the revolutionary democrats. Dmitriĭ Ivanovich Pisarev wrote:

> If the ABC's of natural science were as widespread as those ABC's by which we learn to read, the number of investigators of nature would most likely be increased many times and their works would become so much more fruitful than now because all the results of investigations would be generalized and applied to life more quickly and fully. Routine and prejudices would perish forever because

> they are now supported by the circumstance that the most simple laws of nature are unknown even to educated society. Finally, the very temper of minds would be made firmer when the natural sciences are included in the fundamentals of general education[8].

Public scientific lectures, which were read by the formost professors of the higher institutes of education in the central cities of Russia, played no small role in the popularization of natural science and in the struggle against idealism and religious prejudices.

All this could not help but exert a decisive influence on the growth of natural science in Russia in the middle of the nineteenth century. Kliment Arkad'evich Timiriazev wrote:

> As soon as the educated strata of the Russian people felt the emancipation from that oppression which indifferently crushed every thought, this thought displayed its creative force in precisely that direction which was its natural inclination and which coincided completely with the problems of the century[9].

Chemistry experienced an especially large growth during this period. In possession of the atomic-molecular theory, Butlerov's theory of chemical structure, and Mendeleev's immortal discovery of the periodic law of the chemical elements, chemistry attained brilliant successes in the field of theory as well as in experimentation and, according to the words of Chernyshevskiĭ, constituted "almost the greatest glory" of the nineteenth century.

Russian scholars played an exceptionally important role in the growth of chemistry. The works of the well-known chemists, N. N. Zinin, A. M. Butlerov, Nikolaĭ Nikolaevich Beketov, Vladimir Vasil'evich Markovnikov, Nikolaĭ Aleksandrovich Menshutkin, and the brilliant theoretical generalizations and experimental researches of D. I. Mendeleev and others constitute the foundation of chemical science. Characterizing the significance of the works of Russian chemists, the President of the British Chemical Society, [William Palmer] Wynne [(1861–1950)], stated:

> If, appraising the musical school connected with the names of Balakirev, Borodin (he is also a chemist), Rimsky-Korsakov, Tchaikovsky; or appraising the writers Turgenev, Dostoevsky, Leo Tolstoy, and their contemporaries, we reckon that without them society would be immeasurably poorer, then it would not be an understatement to say that the growth of chemistry to no small degree would be delayed if the works of Mendeleev, Butlerov, Wagner, and their successors for some reason were removed from the general treasure house of knowledge[10].

Not only chemistry attained brilliant successes in Russia during this period. The classical works of the physiologist Ivan Mikhaĭlovich Sechenov, of the botanists Lev Semënovich Tsenkovskiĭ, Andreĭ Sergeevich Famintsyn, and Andreĭ Nikolaevich Beketov, of the embryologists Aleksandr Onufrievich Kovalevskiĭ and Il'ia Il'ich Mechnikov, of the paleontologist Vladimir Onufrievich Kovalevskiĭ, of the mathematician Pafnutiĭ L'vovich Chebyshev, and many others advanced Russian science to one of the top places in the world.

The characteristic trait of the leading men of Russian natural science was the fact that they did not restrict themselves to abstract and laboratory experimental work. They gave all their energy and knowledge to their native land and its people. They labored for its enlightenment, assisting the growth of Russia's industrial powers.

Aleksandr Porfir'evich Borodin belongs to this remarkable galaxy of Russian scholar-patriots. He was an outstanding organic chemist, making a rich contribution to the storehouse of chemistry, and at the same time he was a composer of genius.

Footnotes to Introduction

1 V[ladimir] I[l'ich] Lenin: Sochinenie (Collected Works). Vol. 16, p. 252. — A.

2 The first modern Russian revolutionaries, who led an unsuccessful uprising on December 26 (December 14 O.S.), 1825 and who, through their martyrdom, provided a source of inspiration to subsequent generations of Russian revolutionaries. — T.

3 October 1853-February 1856, fought mainly on the Crimean Peninsula between the Russians and the British, French, and Ottoman Turks, with support (from January 1855) by the army of Sardinia-Piedmont. — T.

4 The organ of local self-government in European Russia and the Ukraine; established in 1864 to provide social and economic services, it became a significant liberal influence within imperial Russia. — T.

5 V[ladimir] I[l'ich] Lenin: Sochinenie (Collected Works). Vol. 20, p. 173. — A.

6 Ibid, p. 174. — A.

7 Ibid, p. 175. — A.

8 D[mitriĭ] I[vanovich] Pisarev: Sobranie sochineniĭ (Collected Works). Vol. 3, 3rd ed., 1900, pp. 109–110. — A.

9 K[liment] A[rkad'evich] Timiriazev: Probuzhdenie estestvoznaniia v tret'eĭ chetverti XIX veka. Istoriia Rossii v XIX v. (The Awakening of Natural Science in the Third Quarter of the 19th Century. History of Russia in the 19th Century) [hereafter abbreviated as Awakening], No. 26, p. 30. — A.

10 [William Palmer] Uinn (Wynne): O znachenii rabot russkikh khimikov dlia mirovoĭ khimii (The Significance of the Works of Russian Chemists for World Chemistry), Leningrad, 1924, pp. 7–8. — A. [For more on the relations between Russia and Great Britain see: USSR Academy of Sciences: Scientific Relations with Great Britain. Nauka Publishing House, Central Department of Oriental Literature, Moscow, 1977 and Yu. I. Solov'ev: D. I. Mendeleev and the English Chemists, Journal of Chemical Education, *61*: 1069–1071 (1984). — T.]

Chapter 1

Childhood and Youth

Aleksandr Porfir'evich Borodin was born on October 31 [according to the Old Style (O.S.); Nov. 12 according to the New Style (N.S.)[1]][2] 1833, in St. Petersburg on Gagarinskaia Street. He was the natural son[3] of Luka Stepanovich Gedianov (1772—1840) and Avdot'ia Konstantinovna Antonova[4]. He received his family name and patronymic[5] from Gedianov's valet, Porfiriĭ Ionovich Borodin, and was recorded as his son in the birth register.[6] His "illegitimate" appearance into society was fortunately not reflected in his education. In spite of her poor education, his intelligent, energetic mother took every measure to ensure that her dear son's childhood should not be clouded. He learned the lessons of primary and intermediate school at home with the help of teachers engaged by his mother.

Borodin showed a love for music at an early age. His wife, Ekaterina Sergeevna Borodina, writes in her memoirs:

> Sasha[7] was 8 years old. Sometimes military music was played at a parade, and Sasha, accompanied by Louise (the nursery governess), set out to listen to it without fail. He was well acquainted with the musicians, he examined their instruments, he watched as they played on each of them. And at home he sat at the piano and played by ear what he had heard. Seeing his love and aptitude for music, his mother arranged for him to take lessons on the flute.

A soldier from the military orchestra of the Semenov Regiment "arrived to teach him for fifty kopecks a lesson"[8].

As a 9-year old, Borodin wrote a polka entitled Hélène, his first musical composition, dedicated to Elena, a young girl with whom he was in love. The gifted youth, who possessed an exceptional memory, easily mastered academic subjects, and in particular he quickly became proficient in German, French, and English. Subsequently he also studied Italian. The 12-year-old Borodin became acquainted with Misha[9] Shchiglev[10], and a warm friendship began between them, which endured throughout their lives.

According to Shchiglev,

> Early on Borodin and I began to enjoy orchestral music. We listened to the orchestra in Pavlovsk[11], where Johann Gungl[12] played at that time. Shortly afterward, symphonic concerts were begun at the university under Carl Schubert's[13] direction. We never missed any of these concerts. In order to become acquainted with chamber music I began to play the violin (self-taught) and Borodin, also self-taught, the 'cello . . .[14].

Fig. 1. Borodin as a youth

Both youths took piano leasons from the music teacher Porman, but his teaching did not satisfy them. To make up for it they played for themselves all of Beethoven's, Haydn's, and Mendelssohn's symphonies in four-hand arrangements with enormous enthusiasm. At that time the young musicians especially liked the sentimental music of Mendelssohn.

In 1847 Borodin composed his first large-scale production, a concerto for flute and piano. He himself played the flute part, and his friend Misha Shchiglev accompanied him on the piano. The same year he wrote a trio for two violins and 'cello on a theme from Meyerbeer's[15] *Robert le Diable* (Robert the Devil). Since he did not attach any great importance to his musical works, Borodin did not write down all of his compositions for a long time. Thus many songs[16] composed during his childhood years are now apparently lost.

However, not only music interested the youthful Borodin. He also studied the natural sciences with no less enthusiasm. Under the guidance of teachers who were invited to his home, Borodin studied basic chemistry with great persistence, trying to understand those phenomena with which he was gradually becoming acquainted. He was preoccupied with the question of why, for example, an ordinary white crystalline substance is exploded by a blow, why two substances, united one with the other, react vigorously with the formation of completely new substances, etc. Shchiglev relates,

> Almost the entire apartment was filled with jars, retorts, and all sorts of chemicals. Tubes with various crystalline solutions[17] were on windows everywhere. And they mildly criticized Sasha Borodin: in the first place, the whole house smelled of his chemical preparations, and secondly, they were afraid of a fire. In the time free from his lessons he studied modeling from wet paper, galvanoplasty,[18] and he designed and prepared water colors[19].

In 1849 his anxious mother, knowing how difficult it is to be from the peasant class, found the money to "register the emancipated serf of Prince Gedianov, the manor serf of Saratovskiĭ Province, Balashovskiĭ District, village of Novoselok, Aleksandr Porfir'evich Borodin, in the Novotorzhskiĭ merchants of the third guild"[20]. In 1850, at the age of 17, Borodin brilliantly passed the entrance exams for the Medical-Surgical Academy and enrolled there as an external student.

Footnotes to Chapter 1

1 Russia continued to use the Julian (Old Style) calendar until February 1, 1918, when the USSR adopted the Gregorian (New Style) calendar. — T.

2 The literature usually cites 1834 as the year of Borodin's birth. In 1927–1929 S.A. Dianin discovered documents in the archives proving that Borodin was born on Oct. 31, 1833. — A.

3 Borodin was not the only illegitimate Russian son to achieve outstanding success in chemistry. For example, another prominent Russian chemist who was also a bastard was Chugaev. See George B. Kauffman: Terpenes to Platinum: The Chemical Career of Lev Aleksandrovich Chugaev. Journal of Chemical Education. *40*: 656–665 (1963) and A Russian Pioneer in Platinum Metal Research: The Life and Work of Lev Aleksandrovich Chugaev. Platinum Metals Review. *17*: 144–148 (1973). — T.

4 Avdot'ia Konstantinovna Antonova (1809–1873) [,the daughter of a soldier,] was born in the town of Narva [about 90 miles – ca. 145 kilometers – west of Leningrad]. She later [Spring 1839] married the military doctor Khristian Ivanovich Kleĭneke. — A.

5 A name derived from that of a father or paternal ancestor, usually by the addition of a suffix or prefix. In Russian society everyone is referred to by a patronymic as well as a first name and surname. Thus Aleksandr Porfir'evich Borodin is Aleksandr, the son of Porfiriĭ Borodin; his mother was Advot'ia Konstantinovna Antonova, i.e., Avdot'ia, the daughter of Konstantin Antonov. Patronymics occur in other languages, e.g., the Scottish name MacDonald originally meant son of Donald, the English name Williamson originally meant son of William, etc. In Iceland patronymics are used instead of family names; thus, husbands and wives as well as their offspring have different last names. — T.

6 Porfiriĭ Ionovich Borodin's wife, Tatiana Grigor'evna Borodina, is listed in the register as Aleksandr's mother, and Sergeĭ Konstantinovich Antonov and Ustinia Konstantinovna Gotovtseva (née Antonova), his maternal uncle and aunt, respectively, were listed as his godparents. Since Aleksandr's legal parents were serfs of his actual father, Aleksandr was legally born as his own father's serf. According to the documents concerning his paternal ancestors, the Gedianov family was of Tartar origin, while Vladimir Vasil'evich Stassov (A. P. Borodin, ego zhizn', perepiska i muzykal'nye stat'i (A. P. Borodin, His Life, Correspondence, and Musical Items) (hereafter abbreviated as Borodin), Souvorin, St. Petersburg, 1889, p. 1) states that Borodin descended "by his father from a family of Princes of Imeretia." Aleksandr and his real father, who strongly resembled each other, possessed clearly Georgian features. Sergeĭ Aleksandrovich Dianin discusses Aleksandr's ancestry in great detail (Borodin. Oxford University Press, London, New York, 1963, pp. 2–8, 344–347 and the genealogical table at the back of the book). Aleksandr's mother never acknowledged her Sasha as her son; she always referred to herself as "Auntie," the name with which he addressed her in all his letters. — T.

7 Sasha is the diminutive form of Aleksandr. — T.

8 V. V. Stassov: Borodin, p. 4. — A.

9 Misha is the diminutive form of Mikhail. — T.

10 Mikhail Romanovich Shchiglev (1834–1903) was the oldest of the Shchiglev brothers, a childhood friend of Borodin's, a teacher of music, and a composer. — A.

11 A town on the southern outskirts of Leningrad (then St. Petersburg). — T.

12 Johann Gungl (1828–1883) was a [Hungarian conductor and] composer who composed primarily dance music. — A. [See Andrew Lamb: Johann Gungl. In: Grove, Vol. 7, p. 848. — T.]

13 Carl Schubert (1811–1863) was a [German] 'cellist and conductor. — A. [See William Barclay Squire: Carl Schubert. In: Grove. Vol. 16, p. 812. — T.]

[14] V. V. Stassov: Borodin. p. 6. — A.

[15] Giacomo Meyerbeer [Jakob Liebmann Meyer Beer] (1791–1864) was a German composer. In 1831 he wrote [his first French] opera Robert le Diable, [which was an immediate success]. — A. [See Heinz Becker: Giacomo Meyerbeer. In: Grove. Vol. 12, pp. 246–256. — T.]

[16] For details on Borodin's songs see Gerald Abraham: Borodin's Songs. Musical Times. *75*: 983–985 (1934). — T.

[17] This apparently refers to solutions from which crystallization had taken place. — A.

[18] The process of making electrotypes. — T.

[19] V. V. Stassov: Borodin. p. 4. — A.

[20] The complete text of this document is published in S[ergeĭ] A[leksandrovich] Dianin: Borodin; zhizneopisanie, materialy i dokumenti, pod. obshcheĭ red. I. F. Belza i V. A. Kiseleva. Gos. muzykal'noe izdatel'stvo, Moscow, 1960 (Borodin, Biography, Materials and Documents), Moscow, 1955; 2nd edition, 1960. — A.

Chapter 2

The Medical-Surgical Academy

The Medical-Surgical Academy of St. Petersburg was established on December 18, 1798. From the first year of its activity it has occupied a most outstanding position among Russian institutions of higher education. During the 1850's and 1860's it was an important center of natural scientific ideas and natural science education. Prominent scholars read lectures there, among whom were Nikolaĭ Nikolaevich Zinin, "The Nestor of Russian Chemists"[1], the famous N. I. Pirogov[2], and Academician F. F. Brandt[3]. Its students made use of a rich library, an excellent botanical garden, the huge collections of the zoological museum, and the academy's mineral collection.

The Chair of Chemistry was one of the first to be organized at the academy's opening. In the first years of its existence, the academy's chemical laboratory had poor equipment. Some primitive apparatus and reagents for experimental lecture demonstrations were all that was available at the time. The department was located in one room on the lower floor of the awkward, two-storied stone building on the bank of the Neva River.[4]

The famous Russian mineralogist and chemist, Academician V. M. Severgin[5], was the first professor to occupy the Chair of Chemistry – from 1799 to 1804. From 1804 A. I. Sherer[6] occupied the chair. He acquired a collection of wax models of crystals for the study of chemistry from the heirs of the famous Russian chemist T. E. Lowitz[7], and he published an original handbook on chemistry in Russian. Nevertheless, the academy's chemical laboratory in general led a miserable existence for many years – until Zinin's appointment to the head Chair of Chemistry. In 1823 Professor S. Ya. Nechaev[8] succeeded him in the Chair of Chemistry.

Nechaev was a highly educated professor. Together with Mikhail Fedorovich Solov'ev and P. Sobolevskiĭ[9], he took an active part in the development of Russian chemical terminology. After the departure from the academy of the famous physicist Vasiliĭ Vladimirovich Petrov, [10] Nechaev was also entrusted with the teaching of physics, but soon (1836) the Chairs of Physics and Chemistry were united. G. H. Hess'[11] rextbook Osnovaniia chistoĭ khimii (*Foundations of Pure Chemistry*), which went through six editions, provided the basic instruction in chemistry during the 1830's and 1840's.

The condition of the chemical laboratory was somewhat improved under Nechaev's direction. The Chemistry Department was outfitted with physical and chemical apparatus which was obtained from Sherer's personal laboratory. Furthermore, in 1837 Nechaev was sent abroad by the Conference of the Academy specifically to acquire apparatus and equipment for the medical and physical-chemical

Fig. 2. St. Petersburg Medical-Surgical Academy

laboratories. However, despite these improvements in equipment, work with the students was conducted irregularly and in a limited way, especially in winter when the temperature in the laboratory fell to 5 °C. (41 °F.) because of its poor location.

Nechaev left the Chair of Chemistry and Physics in 1847, and the chemistry professor at Kazan University, Nikolaĭ Nikolaevich Zinin, Borodin's future teacher, was appointed to succeed him.

Zinin was born on August 13, 1812 in [Shusha,] Transcaucasia [now the Azerbaijan Soviet Socialist Republic — T.]. Orphaned in his early years, he was forced to make his way in the world on his own. Upon graduating from the Saratovskiĭ Gymnasium[12], he enrolled at Kazan[13] University in 1830 in the Mathematics Department of the Physical-Mathematical (Philosophical) Faculty. Thanks to his brilliant abilities, he soon stood out among his schoolmates and brought himself to the attention of the professors, in particular, that of the famous mathematician, N. I. Lobachevskiĭ[14]. After brilliantly completing his studies at the university in 1833, Zinin remained as a tutor in the Department of Physics and began his own teaching career. In 1835 he was entrusted with the instruction in pure chemistry. In the following year he defended his Master's dissertation: O avleniiakh khimicheskogo srodstva i o prevoskhodstve teorii Bertseliusa o postoiannykh khimicheskikh proportsiiakh pered khimicheskoiu statikoiu Bertolleta (On the Phenomena of Chemical Affinity and the Superiority of Berzelius' Theory Concerning Constant Chemical Proportions over Berthollet's Chemical Statics).

After defending his dissertation, Zinin was sent abroad and stayed in various countries for almost 3 years. Returning to Kazan, he developed a broad scientific activity. In 1842 he discovered the reaction for the conversion of aromatic nitro compounds into amines (aniline, naphthylamine, etc.), which earned him worldwide fame.

Several years before this discovery, the Russian Academician Yu[liĭ] F[edorovich] Fritsche[15] had obtained aniline by the decomposition of the expensive vegetable dye indigo, thus demonstrating that a molecule of this dye is composed of an aniline group. Zinin's discovery of this easy method for obtaining cheap aniline was the first and important step in the synthesis of dyes.

Fig. 3. Nikolaĭ Nikolaevich Zinin

Zinin's reaction laid the foundation for the production of dyes and initiated the inception and vigorous growth of the aniline dye industry. Up to the present time aniline, obtained by Zinin's method, serves as the most important product for the synthesis of various organic substances and is prepared in large quantities in the chemical industry.

With Zinin's arrival at the Medical-Surgical Academy in 1847 the condition of the teaching of chemistry was improved. Being a highly educated scholar who brilliantly understood the significance of the natural sciences in medical education, Zinin quickly introduced various improvements in teaching. In contrast to his predecessors, he took upon himself the teaching of chemistry only; physics was assigned to the Ad'iunkt [Assistant] A. Izmailov. At Zinin's suggestion, inorganic and analytical chemistry were read for the first of the academy's courses; two lectures per week were read in inorganic and three in analytical chemistry, each [lecture being] one and a half hours long. Organic chemistry with its application to physiology was read for the second course.

Characterizing Zinin's activity as a professor at the academy Borodin wrote:

> Being one of the professors of the St. Petersburg Medical Academy, N. N. Zinin transferred here those lively and lofty principles of rigorous science, progress, and independence, which he formerly championed in Kazan. His word from the chair was not only a reliable reproduction of contemporary conditions but also a tribune of new trends in science. He was not sparing in ideas; he cast them to the right and to the left, and time after time he developed ideas in lectures which were only heard of several years later, such as a new discovery or a new scientific idea. . . . Beginning scholars and visitors hurried to communicate the results of their first works or to consult with the authoritative experienced master of the Medical-Surgical Academy, Zinin, about their ideas, plans, [and] intentions.

... The laboratory was converted into a miniature chemical club, into an improvised session of the Chemical Society where the life of young Russian chemistry bubbled up . . .[16]

Considering chemistry as one of the basic sciences necessary for the medical doctor, Zinin noted that

medicine, like science, represents only an application of natural science to the problem of the preservation and restoration of health. Natural science therefore must play a most important role in medical education, but by no means must it be a supplementary or an auxiliary aid. The physician needs to master general systems of science, the method of thinking, [and] devices and methods of investigation rather than many fragmentary facts of applied natural sciences. Therefore the teaching of the natural sciences in the academy must be basic and necessarily full, not restricted to the cramped limits of applied knowledge. And for sound learning and a true evaluation of what has been done in science by others, for a clear knowledge of which path science grows by and [how] scientific data are cultivated and increased, it is necessary for each student to work immediately at something independently and, according to his specialty, in some branch of natural science. Only physics and chemistry hold the key to the explanation of all those complex and infinitely different physiological and pathological processes which occur in an organism.[17]

First of all, Zinin supervised the organization of practical study in chemistry for the students. Until his arrival at St. Petersburg the chemical laboratory was, as we have already said, badly organized, although it did have some good apparatus and a collection of preparations in the Chemistry Department.

Borodin later wrote about working conditions in the chemical laboratory at the academy:

Conditions in the Department of Chemistry were at that time in a very sad state. Thirty rubles a year were appropriated to chemistry, with the right to ask as much again during the course of the year. We add, that this was the time when it was sometimes impossible to find a test tube on the market in St. Petersburg, when you yourself had to make rubber connections, etc. The academy laboratory offered two dirty, gloomy rooms with arches, stone floors, several tables, and empty cupboards. Because of the lack of hoods, distillations, evaporations, etc. had to be performed in the courtyard, even in winter. Organized practical work was out of the question. But even under these conditions N. N. [Zinin] always found a love for work. Five to six fellows always worked, partly on their own and partly under N. N.'s personal supervision. This situation continued until the beginning of the 60's.[18]

In the history of the Medical-Surgical Academy, Zinin's tenure was marked by various innovations which raised the instruction to a high level and made the academy chemistry laboratory one of the most prominent scientific centers in the country. Zinin succeeded in bringing about the construction of a special building of the academy called the Estestvenno-Istoricheskiĭ Institut (Natural Science History Institute). At his urgent request new Chairs for Chemistry, Physics, Comparative Anatomy, and Botany were established in the academy in place of the two former chairs, the combined Chair for Chemistry and Physics, and the Chair for Natural Science History.

Zinin obtained an appropriation for 45,000 rubles for the interior arrangement and equipment of the chemical laboratory of the Natural Science History Institute and yearly grants of 2,000 rubles for apparatus and reagents. The institute was opened on October 13, 1863. Its laboratories were equipped in compliance with the latest demands of science. The laboratory stations permitted wide application of practical studies and performance of scientific investigations.

The growth of experimental work at the academy was favored since its newly opened, special departments were filled by first-class scholars, among whom I. M. Sechenov,[19] S. P. Botkin,[20] and I. M. Balinskiĭ[21] must be mentioned.

Borodin was accepted into the Medical-Surgical Academy in 1850, long before the described reforms which immeasurably improved the teaching and scientific work. Borodin's mother was very pleased by her son's entrance into the academy and began to promote his success in every possible way. She moved to a new apartment close to the academy, seeking to create the best possible conditions for her son's schooling. Borodin organized a home chemical laboratory in their new apartment in the Vyborg District [of St. Petersburg — T.] and subsequently carried out a successful series of organic syntheses.

In the academy Borodin studied with youthful ardor the natural sciences that were taught in the first courses. He learned easily and was ahead of those of his age in all the academic courses.

A. P. Dianin[22] wrote,

> For his inquisitiveness in the field of anatomy Borodin almost paid with his life, contracting an infection from a cadaver, from which he recovered only with difficulty.[23]

However, even in the first course, Borodin's main enthusiasm was chemistry. He listened attentively to his famous teacher's fascinating lectures and occupied himself in his home laboratory, since student participation in practical chemistry work did not yet exist in the academy at that time.

During the third course, after much hesitation, Borodin decided to apply to Zinin, requesting permission to work in the professor's chemical laboratory under his guidance. This was the first time in the academy that Zinin had ever received such a request from a student. At first he doubted the young man's wish, since he was not certain that a medical student would be seriously interested in chemistry. However, after convincing himself of Borodin's abilities and of the sincerity of his intention, Zinin agreed to comply with his request, and Borodin had the opportunity of working in the laboratory of one of the world's most important chemists even before finishing his courses at the academy. After only a few months, Zinin knew that he had not made a mistake in allowing Borodin in his laboratory. The young man showed promise of becoming, in time, an outstanding scholar, and Zinin soon publicly announced in his lectures that he looked upon Borodin as his successor.

Borodin's enthusiasm for music somewhat troubled Zinin. According to a recollection of A. P. Dobroslavin's,[24] Borodin, when working in the laboratory, "almost always hummed to himself." At home Borodin played the piano and systematically listened to concerts and operas during all his free time. Zinin often said to Borodin:

> Mr Borodin, concern yourself a little less with songs. I am pinning all my hopes on you to prepare you as my successor, and you think of nothing but music; you cannot hunt two hares at the same time[25].

However, in Borodin's words, he did not think about "two hares" at all. In letters to his friends and acquaintances he emphasized that his musical enthusiasm was only "relaxation".

Meanwhile, during the years 1854–1855 Borodin zealously worked on mastering the technique of composition and composed many fugues during this period. According to Shchiglev,

> We miss no opportunity for playing through a trio or a quartet wherever and with whomever. Neither bad weather, rain, nor slush, nothing held us back. I, with my violin under my arm, and Borodin, with his 'cello in a blanket knapsack on his back, sometimes went great distances, for example, from Vyborg to Kolomna,[26] on foot, since we didn't have any money, not even half a kopeck[27,28].

Becoming closely acquainted with I. I. Gavrushkevich's musical circle, both friends took pleasure in the performance of the fashionable chamber music of Mendelssohn, Gebel[29], Boccherini[30], and Onslow[31]. Gavrushkevich noted,

> Borodin listened to Boccherini's quintet with curiosity and youthful sensitiveness, to Onslow with pleasure, and to Gebel with love. He detected the influence of Russian Moscow in Gebel. Germans did not like this German because he reeked of Russia.[32]

Thirty years later Borodin wrote to Gavrushkevich,

> Very often and very warmly I remember you and our evenings, which I loved so and which were a serious and good school for me, as serious chamber music always is.[33]

A. N. Serov,[34] sometimes attending Gavrushkevich's quartet meetings, captivated the youth by his execution of the compositions of Glinka, the brilliant Russian composer. Glinka's rich, deeply national music fascinated and captivated the youthful musician [Borodin. — T.]. Subsequently he always rated Glinka's works highly. In a letter to his mother dated November 5, 1859 from Heidelberg Borodin wrote:

> In addition, he [I. G. Borshchov — A.] turned out to be a very good musician with "*our*" inclinations in music. I was very happy because of this meeting . . .

(It is necessary to note that Borshchov[35] is a zealous admirer of Glinka and knows his operas by heart from cover to cover) . . .[36]

It is interesting to note that Borodin wrote this letter before meeting M. A. Balakirev[37], who later played an important role in the formation of Borodin's creative aspect as a composer.

In 1855, during the period of preparing for final examinations at the academy, Borodin composed a trio for two violins and 'cello on the theme of an ancient Russian song Chem tebia ya ogorchila (How Have I Offended Thee), the song Krasavitsa Rybachka (The Beautiful Fisher Maiden), and a piano scherzo, in which, in Shchiglev's words, "Borodin's Russian style was encountered for the first time"[38].

Work in Zinin's laboratory proceeded successfully, although, according to Dianin's recollections,

> the new laboratory, in which Aleksandr Porfir'evich now worked, was not much different from his home quasi-laboratory. There, too, a lack of vessels, materials, and apparatus constantly hindered work[39].

Nevertheless, work under Zinin's guidance contributed to a great degree to the formation of the future scholar.

Borodin characterized Zinin as a professor and supervisor as follows:

> In addition to the large amount of compulsory work, he always found time to read and to keep up with, not only his own specialty but also developments in the most varied branches of knowledge, current literature, public life, etc., and he also had time to spare for anyone who needed him. And who didn't need him? Thanks to his vast intelligence and phenomenal memory, he was a living, walking reference encylopedia on every branch of knowledge. They turned to him for information about new discoveries in the fields of chemistry, physics, technology, pharmacy, physiology, comparative anatomy, mineralogy, etc.; for suggestions about literature sources for various scientific questions; for an explanation of misunderstandings and contradictions in the scientific literature; for themes for dissertations and scholarly works; for practical advice on how to avoid the difficulty of getting one or another product or to operate some new piece of apparatus; and finally, even for instructions, on how to make an injection of some crayfish, lizard, or turtle, etc. They went to him for appraising the value of a new book when they didn't have time to read it themselves, knowing full well that N. N. had already taken time to look at it thoroughly. Finally, they went to him for advice on everyday questions when it was necessary to come to the assistance of a poor student or medical doctor who was worried by need or upon whom some misfortune fell, in other words, when a man needed help, moral or mental.
>
> Kind to the highest degree, humane, accessible to each and every person, always ready to help by both words and deeds, N. N. never refused anyone. His warm interest in people, his desire and ability to help everyone, to be of any possible use, his extreme simplicity in treatment, affability, and cordiality soon

> made his name one of the most popular in the Medical-Surgical Academy. He was astonishingly able to inspire faith, love, and respect. But if personal qualities created ardent admirers and devoted friends for him, personal qualities also made him not a few enemies, as a person and as a public figure. His passionate and ardent nature could not tolerate triteness, vanity, ignorance, or lack of talent. He would not endure anything routine or petty, either in science or in life. His penetrating mind soon detected these elements, no matter how skillfully they were disguised or how they were masked by authority. With witty sarcasm he pointedly and mercilessly branded them whenever he encountered them. With a single word he was sometimes able to disperse the thick mist of false learning and to expose the lack of talent and ignorance hidden under it. Their pride outraged, their god and high priest dethroned, they of course could never be able to forgive him for this, and they took their revenge at every opportunity.
>
> A fervent patriot, who loved Russia deeply but judiciously, who understood and took its interests to heart, N. N. first of all zealously advocated the autonomy of Russian science and the intellectual growth of the Russian man. In conflict with administrative and public elements, whose personal sympathies or interests were contrary to his leanings, he had to enter into battle for the principles dear to him. Conqueror or conquered, with a shield or on a shield, he nevertheless made irreconcilable enemies for himself. And he was so hated by them that he belonged to an unusual group of warriors. Generously gifted with the natural qualities of a lively, clear mind, resourcefulness, quick consideration, passion, and energy and equipped with complete knowledge, experience, and brilliant dialectics, he always represented a dangerous opponent. . . . His enemies were not always scrupulous as to their means and did not hesitate to slander Zinin's lucid personality. Alien to every [kind of] vindictiveness, gentle and good-natured as are the majority of good Russian people, N. N. regarded this indifferently enough and confined himself largely to laughter, a joke, or a well-aimed witty remark.[40]

While a student at the Medical-Surgical Academy, Borodin was neither for nor against the public movement which captivated the student body. The student youth of the academy had every opportunity to absorb greedily the creations of the outstanding representatives of the Russian intelligentsia, Belinskiĭ, Dobroliubov, and Chernyshevskiĭ.

There were fresh recollections of the revolutionary events of 1848 in Western Europe in the student environment of the Medical-Surgical Academy during the years when Borodin studied there (1850–1856). Student groups discussed the questions of the slave position of the Russian peasants, of the lack of rights of the people, and of the arbitrary rule of the Czarist officials. Such student circles during the middle of the last century were the germs of the future revolutionary unions.

Nicholas I,[41] already frightened by the uprising of the Decembrists[42] in 1825 and by the revolutionary movement in Western Europe during the years 1848 and 1849, thought that by introducing a very severe control over the "intellect" of the students, by restricting schools of higher education from penetration by any revolutionary "spirit", and by the strictest repressions, he could stop the spread of "sedition". One of these measures was restricting the admission of *raznochinets*[43] to the university. But none of the Czarist autocracy's efforts which were directed to the suppression of "free thinking" accomplished its goal. Russia's *raznochinets* students remained revolutionaries and exerted a great influence on the spread of revolutionary ideas among Russia's intelligentsia.

Sechenov, while a student at Moscow University, wrote of these times:

> It is well known that when the revolutionary movement of 1848 and 1849 approached our borders from Prussia and Austria, Emperor Nicholas found it necessary to take special measures against the penetration of harmful ideas from the West, and one of these measures was the curtailment in Moscow University (I do not know if this measure also spread to other universities) of the number of students in all faculties except medicine to three hundred.[44]

Footnotes to Chapter 2

1 See footnote[42], Chapter 10. — T.

2 Nikolaĭ Ivanovich Pirogov (1810–1881) was a famous Russian doctor and teacher. Born in Moscow, [as a precocious youth,] he graduated from Moscow University. From 1841 [, at the early age of 30,] he was professor of hospital surgery and applied anatomy in the Medical-Surgical Academy. [He retained this chair for 16 years, whereupon he resigned in disgust.] He was the first to use anesthesia widely for surgical operations. — A. [For a short treatment of his life and work and pertinent references see S. R. Mikulinsky: Nikolay Ivanovich Pirogov. In: Charles Coulston Gillispie (ed.): Dictionary of Scientifc Biography [hereafter abbreviated as DSB], Charles Scribner's Sons, New York, 1974, Vol. 10, pp. 619–621. — T.]

3 Fedor Fedorovich [Johann Friedrich] Brandt (1802–1879) was a zoologist. An academician from 1833, he occupied the Chair of Zoology and Comparative Anatomy at the Medical-Surgical Academy. — A. [See Great Soviet Encyclopedia (English translation of the 3rd edition of A. M. Prokhorov (Editor in Chief), Bol'shaia Sovetskaia Entsiklopediia), 30 vols., Macmillan, Inc., New York, 1973–1982 [hereafter abbreviated GSE], 1973, Vol. 3, p. 523. — T.]

4 The Neva River, the outlet for Lake Ladoga, from which it issues via a delta into the Gulf of Finland, flows through the city of St. Petersburg (now Leningrad). — T.

5 Vasiliĭ Mikhaĭlovich Severgin (1765–1826) was born in St. Petersburg. He occupied the Chair of Chemistry at the Medical-Surgical Academy from 1799 to 1804. From 1793 he was Academician of Mineralogy. — A. [For further information on his life and work and pertinent references see G. D. Kurochkin: Vasily Mikhaylovich Severgin. In: DSB, 1975, Vol. 12, pp. 329—330. — T.]

6 Aleksandr Ivanovich Sherer (1771–1824). Born in St. Petersburg, he wrote the well-known textbook Rukovodstvo k prepodavaniiu khimii (Handbook of the Teaching of Chemistry), 1808. He was an academician from 1815. — A.

7 Toviĭ Egorovich Lovits (Tobias Lowitz) (1757–1804) was an outstanding Russian chemist and academician. — A. [See N. Figurovsky: Johann Tobias Lovits (Lowitz). In: DSB, 1973, Vol. 8, pp. 519–520. — T.]

8 Stepan Yakovlevich Nechaev (1799–1862) graduated from the Medical-Surgical Academy in 1818. From 1823 he was Professor of Chemistry at the academy. — A.

9 Pëtr Grigor'evich Sobolevskiĭ (1782–1841), industrial engineer and metallurgist, founder of powder metallurgy. — T.

10 Vasiliĭ Vladimirovich Petrov (1761–1834) was a well-known Russian electrotechnician and experimental physicist. He discovered the phenomenon of the electrical arc light (1803). From 1793 to 1833 he occupied the Chair of Physics at the Medical-Surgical Academy. — A. [See Olga A. Lezhneva: Vasily Vladimirovich Petrov. In: DSB, 1974, Vol. 10, pp. 552–553. — T.]

11 German Ivanovich Gess (Germain Henri Hess) (1802–1850) was an outstanding chemist and the founder of thermochemistry. He was professor at the Main Pedagogical Institute at St. Petersburg (1832–1848). The first edition of Hess' textbook appeared in 1833, the sixth edition in 1845. — A. [See Henry M. Leicester: Germain Henri Hess. In: DSB, 1972, Vol. 6, pp. 353–354. — T.]

12 In Europe, a classical school preparatory to the university, roughly equivalent to an American high school. — T.

13 City about 450 miles (ca. 725 km) east of Moscow, on the eastern shore of the upper part of the Kuybyshev Reservoir on the Volga River, where it is joined by the Kazanka River. Now the capital of the Tatar Autonomous Soviet Socialist Republic, its university was attended by Tolstoy, Balakirev, and Lenin. — T.

14 Nikolaĭ Ivanovich Lobachevskiĭ (1793–1856), a great Russian mathematician, was the creator of non-Euclidean geometry. In 1827 he was elected rector of Kazan University and occupied this position for 19 years. He was subsequently assistant trustee of the Kazan district. — A. [He was satirized in a song by Tom Lehrer, recorded in the long-playing record album, Songs by Tom Lehrer (His Lyrics, His Music, His So-Called Voice, and His Piano). Reprise RS-6216. See also B. A. Rosenfeld: Nikolai Ivanovich Lobachevsky. In: DSB, 1973, Vol. 8, pp. 428–435. — T.]

15 Known in the West as Carl Julius Fritsche, he was born in Neustadt bei Stolpen, Saxony in 1808 and died in Dresden in 1871. An assistant to Eilhard Mitscherlich (of isomorphism fame), he migrated to Russia in 1833 and became a member of the St. Petersburg Academy of Sciences. He rediscovered Crystallin, first discovered and named in 1826 by Otto Unverdorben, in 1840 and renamed it from the Arabic, al-nil (indigo). — T.

16 Zhurnal Russkogo fisiko-khimicheskogo obshchestva (Journal of the Russian Physical-Chemical Society). *12* [(5): 215–254 (1880)]. — A. [This is Borodin's Publication No. 37 in the Complete List of Borodin's Chemical Works, which begins on page *121*. Numbers prefixed with the letter *B* refer to Borodin's publications in this list. — T.]

17 Istoriia Mediko-Khirurgicheskoĭ Akademii za sto let, 1798–1898 (History of the Medical-Surgical Academy During the Hundred Years, 1798–1898), St. Petersburg, 1898, p. 529. — A.

18 B[oris] N[ikolaevich] Menshutkin: Nikolaĭ Nikolaevich Zinin. Ego zhizn' i nauchnaia deiatel'nost' (Nikolaĭ Nikolaevich Zinin. His Life and Scientific Activity) [hereafter abbreviated as Zinin], 1921, pp. 58–59. — A. [See also Henry M. Leicester: N. N. Zinin, An Early Russian Chemist. Journal of Chemical Education. *17*: 303–305 (1940); G. V. Bykov: Nikolay Nikolaevich Zinin. In: DSB, 1976, Vol. 14, p. 622, and *B 37*. — T.]

19 Ivan Mikhaĭlovich Sechenov (1829–1905) was a great Russian physiologist. He worked in the field of the physiology of the nervous system. He discovered and described the phenomenon of central inhibition. From 1860 to 1870 he was professor at the Medical-Surgical Academy in the department of physiology. — A. [See M. G. Laroshevsky: Ivan Mikhaylovich Sechenov. In: DSB, 1975, Vol. 12, pp. 270–271. — T.]

20 Sergeĭ Petrovich Botkin (1832–1889) was a well-known Russian therapeutist. He organized the first clinical laboratory in Russia. From 1860 to the end of his life he was professor at the Medical-Surgical Academy in the Department of the Therapeutic Clinic. — A. [See GSE, 1973, Vol. 3, pp. 502–503; 1977, Vol. 15, p. 644. — T.]

21 Ivan Mikhaĭlovich Balinskiĭ (1827–1902) was a psychiatrist. He organized the first clinic for mental discorders in Russia at the Medical-Surgical Academy. From 1860 to 1884 he was professor at the Medical-Surgical Academy in the Department of Nervous and Mental Disorders. — A. [See GSE, 1973, Vol. 2, p. 584; 1979, Vol. 22, p. 430. — T.]

22 Aleksandr Pavlovich Dianin (1851–1918) was Borodin's successor in the Department of Chemistry at the Medical-Surgical Academy. Entering there in the autumn of 1870, Dianin, influenced by Borodin's and Zinin's lectures, embraced chemistry as his specialty and became one of the most intimate of Borodin's students. In the autumn of 1873 Dianin became Borodin's part-time assistant. In 1880 he was appointed assistant professor in the Department of Inorganic Chemistry. After Borodin's death Dianin read the courses on inorganic, organic, and analytical chemistry, and he conducted practical studies in qualitative and quantitative analysis with the students. In 1895 Dianin became the secretary of the Conference of Medical-Surgical Academies. — A. [See GSE, 1975, Vol. 8, p. 197. — T.]

23 A[leksandr] P[avlovich] Dianin: A. P. Borodin. Biograficheskiĭ ocherk i vospominaniia (A. P. Borodin. Biographical Sketch and Reminiscences) [hereafter abbreviated as Borodin], Zhurnal Russkogo fisiko-khimicheskogo obshchestva (Journal of the Russian Physico-Chemical Society). *20* [4]: 368 (1888). — A.

24 Alekseĭ Petrovich Dobroslavin (1842–1889) was a professor at the Medical-Surgical Academy in the Department of Hygiene and a friend of Borodin's. From 1876 to 1885 he was elected

secretary of the Confederation of the Academies. — A. [See GSE, 1975, Vol. 8, p. 332; 1975, Vol. 6, pp. 691, 692. — T.]

25 V. V. Stassov: Borodin. p. 11. — A.

26 The distance from Vyborg to Kolomna, both districts of St. Petersburg (now Leningrad), is about 3 miles (ca. 5 km). — T.

27 A Russian coin of the smallest denomination, equal to one-hundredth of a ruble, roughly equivalent to an American penny or cent. — T.

28 V. V. Stassov: Borodin. p. 9. — A.

29 Franz Gebel [or Gobel] (1787–1843) was a German composer [and teacher]. In 1817 he moved to Russia. — A. [See Ernst Stöckl: Franz Xaver Gebel. In: Grove, Vol. 7, p. 212. — T.]

30 Luigi Boccherini (1743–1805) was an Italian composer who composed primarily chamber music. — A. [See Stanley Sadie: (Ridolfo) Luigi Boccherini. In: Grove, Vol. 2, pp. 825–831. — T.]

31 Georges Onslow (1784–1853) was a [French] composer [of English descent]. — A. [See Benedict Sarnacker: (André) Georges (Louis) Onslow. In: Grove, Vol. 13, pp. 543–544. — T.]

32 V. V. Stassov: Biografiia A. P. Borodina (Biography of A. P. Borodin) [hereafter abbreviated as Biography]. Istoricheskiĭ vestnik (Historical Herald), St. Petersburg, 1887, p. 143. — A.

33 Ibid. — A.

34 Aleksandr Nikolaevich Serov (1820–1871) was an outstanding Russian composer and musical critic. — A [See Edward Garden: Alexander Nikolayevich Serov. In: Grove, Vol. 17, pp. 178–181. — T.]

35 Il'ia Grigor'evich Borshchov (1833–1878) was a botanist, a professor at Kiev University, and one of the first investigators of colloid chemistry. — A. [See GSE, 1973, Vol. 3, p. 482. — T.]

36 Pis'ma A. P. Borodina (Letters of A. P. Borodin) [hereafter abbreviated as Letters, I], Moscow, 1928, pp. 30–31 and 34. — A.

37 Miliĭ Alekseevich Balakirev (1836–1910) was a well-known Russian composer, an outstanding musical figure, the head of the Balakirev circle, well known under the name Moguchaia Kuchka (The Mighty Little Group), and a friend and teacher of Borodin's. — A. [See Gerald Abraham and Edward Garden: Mily Alexeyevich Balakirev. In: Grove, Vol. 2, pp. 47–56. — T.]

38 V. V. Stassov: Biography. pp. 144–145. — A.

39 A. P. Dianin: Borodin. p. 369. — A.

40 B. N. Menshutkin: Zinin. pp. 86–89. — A.

41 Nikolaĭ Pavlovich Romanov (1796–1855), Czar of Russia (1825–1855), was a reactionary ruler considered the personification of classic autocracy. — T.

42 See footnote 2 in the Introduction. — T.

43 Intellectuals not belonging to the nobility in nineteenth-century Russia. — T.

44 Ivan Mikhaĭlovich Sechenov: Avtobiograficheskie zapiski (Autobiographical Notes), [Izdatel'stvo Nauchnavo slova], 1907, p. 70. — A.

Chapter 3

The Young Scientist's First Steps

In 1856 the student Aleksandr Borodin passed his final examination with distinction and graduated from the academy with the school testimonial of good conduct and progress. Although he ranked first in his course, the reason that he did not receive the Gold Medal was a bad grade received on a junior course given by the teacher of theology, Cherepnin, because he had explained a text of Holy Writ freely, whereas its literal translation was needed.

Borodin's outstanding abilities could not help but attract the attention of his teachers. As early as March of 1856, even before the formal completion of [his courses at] the academy, the Professor of General Pathology, N. F. Zdekauer[1], asked the Conference to appoint Dr. Borodin, who distinguished himself "by a particular love for the sciences and who possessed excellent talents," as his assistant. The Conference of the Academy supported this request, and in October of the same year the Military Ministry gave the order for Borodin's affiliation as intern at the Second Military-Land Forces Hospital.

The brilliant prospect of becoming in time a professor of medicine was opened to Borodin. However, the work of a medical doctor not only did not attract him but also did not conform to his character. Once, for example, he had to extract splinters from the backs of six bonded peasants who had "run the gaunlet." Borodin fainted three times at the sight of the blood and the skin hanging loose on their unfortunate backs. The hospital's chief doctor, well acquainted with the duties of a medical doctor in Czarist Russia, remarked upon this occasion: "Eh, young fellow, what will you do if in the performance of your duty you are requested to lay the branding iron on a convict?"[2]

In September, 1856 Borodin began the session of examinations for the degree of Doctor of Medicine. The examinations consisted of a theoretical part, practical tests, and the solution of written problems. He passed theory, especially therapeutics, "satisfactorily with distinction"; he passed materia medica[3] and forensic medicine "completely satisfactorily"; but he passed physiology, pathology, general therapeutics, surgery, and obstetrics "satisfactorily". In the practical tests his knowledge of pharmacognosy and pharmacy were judged "satisfactory with distinction" and his knowledge of physiological anatomy, clinic, operative surgery, obstetrics, and dissection of the body "satisfactory". After the successful conclusion of numerous examinations the Conference authorized Borodin to write a doctoral dissertation on the theme: Ob analogii mysh'iakovoĭ kisloty s fosfornoiu v khimicheskom i toksikologicheskom otnosheniiakh (On the Analogy of Arsenic Acid with Phosphoric Acid in Chemical and Toxicological Behavior) [*B2*].[4]

May 3, 1858 was a remarkable day in Borodin's life. After the successful defense of his dissertation, he was awarded the academic degree of Doctor of Medicine. For the first time in the history of the Medical-Surgical Academy a dissertation was written and defended in the Russian language.[5]

In this dissertation, which was the first serious and independent scientific work of the young scientist, Borodin showed himself to be a firm supporter of the new unitary chemical theory of Laurent[6] and Gerhardt[7]. He criticized the dualistic[8] presentation of Berzelius[9] and his followers. The experimental part of the dissertation was devoted to the toxicology of arsenic and phosphoric acids. On the basis of his experiments, Borodin concluded that these acids are analogous in chemical and toxicological behavior and, in contrast to sulfuric, hydrochloric, and nitric acids, do not possess local action.

His preparation for examinations, his strenuous work on his doctoral dissertation, and his occupation in Zinin's chemical laboratory did not prevent Borodin from studying music. He attended concerts and played the piano in his free time. At this time Borodin became acquainted with the 17-year-old officer M. P. Mussorgsky[10]. Here is how Borodin described this meeting:

> My first meeting with Modest Petrovich Mussorgsky was in 1856 (I believe in the autumn, in September or in October). I was an inexperienced military medical man and was an intern at the Second Military-Land Forces Hospital. M. P. was an officer of the Preobrazhenskiĭ Regiment[11] who was just hatched from the egg (at the time he was 17 years old). Our first meeting was in the hospital, in the duty room. I was the duty doctor, he the duty officer. The room was commonplace; it was boring to both of us on duty. It is natural that we got into a conversation and became friends very rapidly because we were both effusive. That same day we were both invited for an evening with Popov, the Chief Doctor of the Hospital. . . . M. P. was at that time quite boyish, very refined, the spit and image of an officer[12].

Fig. 4. Soligalich. View of the health resort (baths) building (1860)

This meeting left a lasting impression on Borodin.

In August, 1857 Borodin was commissioned by the Conference of the Academy to accompany the oculist in ordinary, I. I. Kabat[13], to the International Congress of Ophthalmologists. On their way to Brussels, where the congress was to take place, they visited Germany and France. Borodin naturally wanted to take advantage of this trip to gain an acquaintance with the chemical laboratories of Western Europe and to establish a personal acquaintance with outstanding chemists. In Paris he expected to meet with the laboratory assistants of Balard[14], Berthelot[15], and other chemists, but, as he writes,

> I was not able to find any chemists. With regret I found out that Berthelot is going to Italy the day after tomorrow. I examined his laboratory today[16].

In one of his letters to his mother Borodin wrote that, taking the opportunity of his free time in Paris, he was occupied with the completion of his "memoir." According to S. A. Dianin's view[17], Borodin had in mind here his first chemical work: Issledovaniia o deĭstvii iodistogo etila na gidrobenzamid i amarin i o konstitutsii etikh soedineniĭ (Investigations of the Action of Ethyl Iodide on Hydrobenzamide and Amarine and the Constitution of These Compounds), which he read upon returning from abroad at a meeting of the Physico-Chemical Section of the Academy of Sciences in St. Petersburg on March 5, 1858 [*B 1*]. Upon returning from this mission Borodin broke once and for all with medical practice. He devoted himself entirely to chemical studies, and for the rest of his life he was associated with the chemical laboratory of the Medical-Surgical Academy.

During the summer of 1858 on the advice of the Military Medical Scholarly Council Borodin was sent to Soligalich[18] for the chemical investigation of the mineral waters in order to determine their medicinal uses. Borodin spent almost 2 months in this small town in Kostroma Province, known from ancient times for its salt springs and saltworks. The results of his research proved to be of great interest. The mineral springs of Soligalich undoubtedly possessed medicinal properties, and this provided grounds for establishing a balneological[19] hospital there. The results of Borodin's research and his conclusions were published in the newspaper Moskovskie Vedomosti (The Moscow News)[20] and later printed as a separate booklet[21] [*B4*]. A few years after Borodin's mission in Soligalich a balneological hospital, which still exists at the present time, was built.

Having returned from Soligalich, Borodin occupied himself with scientific-pedagogical work at the academy. During the winter of 1859 he received a commission to conduct the practical study of chemistry for students of the academy's second course and to read lectures to young medical doctors who remained at the academy for advanced courses. Borodin read to the doctors a course on theoretical chemistry as applied to physiology and pathology and, after the example of his teacher Zinin, a course on the history of the development of chemical theories[22]. Borodin's work was highly successful; he devoted his entire energy to his beloved science and teaching.

As before, Borodin was occupied with music only in his free time, when he was resting from his scientific work. The outstanding event in his life during this period was his second meeting with Mussorgsky. This meeting took place in the autumn of 1859. By this time Mussorgsky had grown up physically and spiritually. He already understood the goal of art as a means of spiritual communication of the people. This development in Mussorgsky's views on music proved to be a serious influence on Borodin.

Borodin relates:

> In the autumn of 1859 I saw him [Mussorgsky. — A.] again at the home of S. A. Ivanovskiĭ, the Assistant Professor of the academy and Director of the Artillery School. Mussorgsky had already retired [from the army]. He had grown in height, had begun to put on weight, and had already lost his officer's manner. He possessed elegance in clothing, in manners, etc., but there was not the slightest nuance of foppishness. We were introduced; however, we immediately recognized one another and recalled the first acquaintance at Popov's home. Mussorgsky explained that he had retired because "he was especially busy with music, and combining military service with art is a complicated matter," etc. The conversation involuntarily turned to music. Mussorgsky was already acquainted with Balakirev, and he was familiar with every musical innovation, about which I did not even have the faintest idea. The Ivanovskiĭs, seeing that we found common ground for conversation, music, suggested that we play Mendelssohn's A Minor Symphony[23] in a four-hand arrangement. M. P. made a wry face and said that he would be very happy as long as they "released him from the Andante,[24] which is completely unsymphonic, but resembles one of the 'Lieder ohne Worte' (Songs without Words) or something like it, arranged for orchestra." We played the first movement and the scherzo.[25] After this Mussorgsky began to talk with delight about Schumann's symphonies, which I didn't know at all at that time. He began to play for me parts of Schumann's E Flat Major Symphony.[26] Coming to the middle section, he gave up, saying: "Well, now the musical mathematics begins." All this was new to me, and I liked it. Seeing that I was very interested, he played for me something that was new to me. By the way, I found out that he himself was writing music. Of course, I became interested; he began to play his scherzo for me . . . Coming to the trio, he muttered through his teeth, "Look, this is oriental," and I was terribly amazed by the experimental music which was unprecedented and new to me. I do not say that it especially pleased me to begin with; rather it puzzled me by its novelty. I admit that his statement that he wanted to devote himself to serious music was met by me with distrust at first, and it appeared to be a bit of bragging; inwardly I chuckled a little at him. But, having become acquainted with his "scherzo," I became thoughtful: to believe or not to believe.[27]

Mussorgsky's statement about his desire to devote himself completely to music made a great impression on Borodin. However, duties at the academy, the lectures to the medical doctors, and the strenuous work in Zinin's laboratory always kept him from music. Shortly after this, new vistas of scientific activity were opened to him.

Footnotes to Chapter 3

1 Nikolaĭ Fedorovich Zdekauer (1815–1898) was a therapeutist and public personage. He worked in the Department of the Hospital Clinic at the Medical-Surgical Academy. — A.

2 Branding with a red hot iron of those condemned to penal servitude was part of the duties of medical workers. If the brand was poorly placed and became obliterated with time, the doctor who made the brand was held accountable. — A.

3 Materia medica — a branch of medical science dealing with the sources, nature, properties, and preparation of drugs (ancient name–pharmacognosy). — A.

4 According to Dr. A. A. Benson of the Scripps Institution of Oceanography, La Jolla, California (letter of January 27, 1987 to George B. Kauffman), whose research has been concerned with the toxicity and biochemistry of arsenic, "Borodin's work surely was instrumental in beginning the efforts which led to Paul Ehrlich's success" (the discovery of salvarsan, the first cure for syphilis). — T.

5 As is well known, until the middle of the last century dissertations in higher medical academic institutions in Russia were presented only in Latin. — A.

6 Auguste Laurent (1808–1853) was a French chemist and professor at the University of Bordeaux. He was the author of the nucleus theory of chemistry. — A. [See Satish C. Kapoor: Auguste (or Augustin) Laurent. In: DSB, 1973, Vol. 8, pp. 54–61. — T.]

7 Charles Frédéric Gerhardt (1816–1856) was a well-known French chemist. He was a professor at Montpellier and the author of the type theory of chemistry. He introduced rational formulae and a new chemical nomenclature. — A. [See M. P. Crosland and J. H. Brooke: Charles Frédéric Gerhardt. In: DSB, 1972, Vol. 5, pp. 369–375. — T.]

8 It is interesting that in one of his "musical" letters to V. V. Stassov, Borodin wrote in 1879: "I cannot endure dualism, either in the form of the dualistic theory in chemistry or in biological studies or in philosophy and psychology . . ." — A.

9 Jöns Jacob Berzelius (1779–1848) was a Swedish chemist. He was the author of the electrochemical (dualistic) theory. For more detail on his dualistic theory see p. 38. — A. [Also see Henry M. Leicester: Jöns Jacob Berzelius. In: DSB, 1970, Vol. 2, pp. 90–97. — T.]

10 Modest Petrovich Mussorgsky (1839–1881) was a Russian composter of genius, a member of the Mighty Little Group, and a close friend of Borodin's. — A. [See Gerald Abraham: Modest Petrovich Musorgsky. In: Grove, Vol. 12, pp. 865–874. — T.]

11 A Russian military unit founded by Czar Peter I (the Great) (1672–1725; reigned 1689–1725) to maintain internal security, which was invested in the guards' chancery in 1689. — T.

12 V. V. Stassov: Borodin, pp. 13–14. — A.

13 Ivan Ivanovich Kabat (1812–1884) was a senior doctor at the Optical Department of the Second Military-Land Forces Hospital. — A.

14 Antoine-Jérôme Balard (1802–1876) was a French chemist who discovered bromine. — A. [See Seymour H. Mauskopf: Antoine Jérôme Balard. In: DSB, 1970, Vol. 2, pp. 416–417. — T.]

15 Pierre Eugène Marcellin Berthelot (1827–1907) was an outstanding French chemist. At the time when Borodin was in Paris, Berthelot worked as Balard's laboratory assistant at the Collège de France. — A. [See M. P. Crosland: Pierre Eugène Marcellin Berthelot. In: DSB, 1970, Vol. 2, pp. 63–72. — T.]

16 Borodin: Letters, I, p. 28. — A.

17 Sergeĭ Aleksandrovich Dianin was A. P. Dianin's son. He prepared for publication and provided a commentary for a four-volume collection of Borodin's letters. — A. [See page 132. — T.]

18 About 380 miles (ca. 612 km) east of St. Petersburg (now Leningrad) and about 280 miles (ca. 450 km) northeast of Moscow. — T.

19 Balneology is the science dealing with the therapeutic effects of baths and bathing. — T.

20 Moskovskie Vedomosti (The Moscow News), No. 130 (1859). — A.

21 A. P. Borodin: Opisanie Soligalicheskikh mineral'nykh vod (Description of the Soligalich Mineral Waters), Moscow, 1859. — A.

22 For the first time in Russia Zinin read a course on the history of chemistry (in the Medical-Surgical Academy) during the 1850's. — A.

23 Symphony No 3, in A Minor, Op. 56 (Scottish). — T.

[24] The slow movement, usually the second or third, of a symphony or sonata. In this case, Borodin probably meant the Adagio (third) movement unless the Mendelssohn symphony was actually the A *Major* Symphony (No. 4, Op. 90, Italian). Another possibility would be the Andante con moto introduction of the first movement of the A Minor Symphony although in the next sentence Borodin implies that he and Mussorgsky played the entire first movement. — T.

[25] A movement of light or playful character, especially the second or third movement of a symphony or sonata. If the symphony referred to is Mendelssohn's third, the scherzo is the second movement. — T.

[26] Symphony No. 3, in E Flat Major, Op. 97 (Rhenish). — T.

[27] V. V. Stassov: Biography, p. 147. — A.

Chapter 4

On Foreign Soil

In the autumn of 1859 the Conference of the Academy, because of his "outstanding successes in the sciences," sent Borodin abroad for a period of 2 years for scientific improvement in chemistry (this period later was extended for an additional year)[1].

Borodin viewed the Conference of the Academy's decision with great satisfaction. The foreign mission gave him the opportunity of thinking over and carrying out the research which he had planned. He was now able to become personally acquainted with famous European scientists and to visit their laboratories. In the letter of instructions given by the Conference to Borodin, Professor Zinin indicated that for a well-grounded study of the methods of chemical research Borodin must "visit laboratories and look at the most noteworthy chemical factories" abroad[2].

Borodin did not go abroad as a student. He was a Doctor of Science, who had read a theoretical course of chemistry, who had finished his studies at a school of higher education, and who had his own published works. He was sent abroad to Western luminaries not to complete his studies but for further independent experimental work in chemistry.

In some books and articles the missions of young Russian scientists abroad were considered as journeys for "ideas" and "themes," as if these were lacking in Russian chemists. Short-term work in the laboratory of some chemist abroad was considered as "training" for the young Russian scholar, and often after such training the Russian chemist was affiliated with one or another school and was mentioned in the chemical literature in connection with the name of the head of this school. This was one of the many paths by which the myth of the supposed lack of originality of Russian chemical science arose.

Meanwhile, the historical facts testify to the complete opposite. By no means did Russian chemists go abroad for new "ideas." They swarmed there because, under the severe conditions of material need in which they had to work in Russia, they did not have sufficient time for scientific research. Furthermore, the poor equipment in their chemical laboratories often hindered our scientists from carrying out their experimental projects. Abroad, however, they were able to devote all their time to scientific work; they were materially well off[3] and free both from anxiety and from formal duties.

It is well known, for example, that Butlerov went abroad with ideas and projects more advanced than those which existed there[4]. Even on his first meetings with Western colleagues, in particular with Kekulé, Butlerov developed his new ideas on the chemical structure of organic substances, which lay at the foundation of contemporary organic chemistry[5].

Another great Russian chemist, Mendeleev, who went to Heidelberg (somewhat earlier than Borodin) with the intention of working in Bunsen's laboratory, soon realized that there was nothing for him to learn there. He set up his own home laboratory and in it carried out work on capillary phenomena that he had conceived much earlier. In a letter to L. N. Shishkov[6] Mendeleev wrote:

> I see that there is nothing of value for me in this laboratory; even the weights are rather bad; but mainly there is no neat, quiet, little corner where one can work with such delicate equipment as capillaries . . . The only interesting aspect of this laboratory, alas, is the school itself: a lot of workers, all of them beginners. Thus I decided to construct everything for myself at home[7].

Borodin, who on his day of arrival in Heidelberg already visited this laboratory, wrote to his mother: "... he [Mendeleev — A.] has a very nice laboratory, clean and even furnished with gas."[8]

Borodin's work abroad was also totally independent and original, although it was carried out in Erlenmeyer's[9] laboratory.[10] Borodin belonged to a group of prominent chemists about whom Mendeleev later wrote in connection with the International Chemical Congress in Karlsruhe:

> At the Congress it was pleasant to see that the new principles, which all the young Russian chemists have followed for a long time, took a strong upper hand over routine ideas, which dominated most of the chemists[11].

On November 1, 1859 Borodin left Russia for the small German town of Heidelberg.[12] On the very first day he met with Mendeleev, Sechenov, Savich[13], and other Russian scholars.

At this time there was a large group of young Russian scientists in Heidelberg who were there with the same goal as Borodin. A few Russian families also found themselves in Heidelberg for treatments. All these people soon became friends and united in a society, known by the name Geĭdel'bergskiĭ kruzhok (Heidelberg Circle).

In addition to Mendeleev, Borodin, Sechenov, and Savich, the members of the Heidelberg Circle consisted of Maĭnov, Olevinskiĭ[14], Eshevskiĭ[15], Passek[16], the writer Marko-Vovchok[17], and others. Young Russian scientists – the medical men Botkin and Yunge[18], the chemists Alekseev[19] and Butlerov; the biologists Kovalevskiĭ[20], Tsenkovskiĭ[21], Famintsyn[22], and Lisenko[23]; and others visited in Heidelberg and were acquainted with the members of the Heidelberg Circle.

The head of the Heidelberg Circle of scientists was Dmitriĭ Ivanovich Mendeleev, with whom Borodin quickly made friends. In a letter to his mother from Heidelberg Borodin wrote:

> Russians are divided into two groups: those who do nothing, that is, the aristocrats Golitsins, Olsufevs, etc., etc., and those who do something, that is, who study. These all stick together and meet for dinners and for evenings. I have briefly met all of them with Mendeleev and Sechenov; the latter is a brilliant gentleman, extraordinarily simple and very capable[24].

Here on foreign soil, the idea of creating a Russian Chemical Society was engendered and firmly established by the young Russian chemists. In essence, the Heidelberg Circle itself already represented in embryo form the scientific society. The young Russian scientists, meeting together on evenings after work, discussed burning questions of the growth of science and the results of their own experimental and theoretical investigations. Their plans for the future were made at these meetings.

Describing to P. P. Alekseev his life in Heidelberg and the Heidelberg Circle of Russian scientists, Borodin reported:

> There are huge numbers of Russians here. The majority of them are young. They are capable people, especially in zoology. This is a good situation. Literary evenings are arranged on Saturdays. The chemical society is based here[25].

Of course, not only scientific problems were discussed at the meetings. Young Russian scientists were also vividly interested in social problems of the day.

Usually, after their scientific work in the laboratory, all the members of the Heidelberg Circle met at the home of T. P. Passek. A. A. Yunge comments about this:

> After walks we sometimes dined with our companions, and evenings, of course, we spent at Tat'iana Petrovna's [Passek], around whom a warm, native atmosphere with a strong Russian smell was always generated. The youths used to criticize the professors and the German students with their duels and scars on their faces, and when a speech turned to our Russian affairs, hot and noisy Russian arguments occurred. They were sometimes interrupted by stories, anecdotes, and recollections. Herzen and Pushkin were there on the table; someone would get up and read a favorite part[26].

These conversations "on freedom", the discussion of disturbing scientific and social-political problems, and the reading of Sovremennik (The Contemporary)[27] and Kolokol (The Bell)[28] made a great impression on the young Russians in that group and on the young Borodin. Here in this circle they discussed the ideas of the great Russian enlighteners and revolutionary democrats Herzen, Belinskiĭ, Chernyshevskiĭ, and Dobroliubov, which were profoundly reflected in all Russian science and life in the 1860's.

After 2 years Borodin wrote to Mendeleev from Paris about this foreign period of his life:

> And I, old man, sometimes strongly remember Heidelberg and our comradeship. Give me, God, such a time for the future. As for the others – I do not know, but I was quite well off with you, and in my turn thanks, many thanks to you for a truly comradely disposition, which, I am confident, has not changed in latitude and longitude by that locality where fate takes us anew[29].

The Russians hardly became acquainted with the German residents. In one of his letters to his mother Borodin wrote:

> The society of Germans is unbearable in the extreme. They are stiff and are horrible gossips. If you are in a home two or three days running where there are grown-up daughters and, good God preserve us, you play duets with them, believe me, than on the next day they will be talking about you as a fiancé ... The society of German students is still more offensive: horrible schoolboy's tricks; they are real children. Imagine! They are divided into parties, each of which has its seignior. Students of different parties are distinguished by costumes and colors; one has yellow service caps, the next red, the third white, etc. In addition to this, each student has a silk belt across the shoulder; the seignior has a three-cornered hat; the style of the service caps is most curious! – Add to this huge jackboots of a very strange design – and you will have an idea of the costume of the German student. On Sundays the students drink heavily, and it is a rare week that passes without a duel; the cause of the duel is always the same: one student calls the other "dummer Junge" (stupid boy). And this has been happening from time immemorial. Here is conservatism![30]

During the winter of 1859–1860 Borodin worked a great deal and "with relish" in the laboratory on investigations of benzidine derivatives. This theme was undoubtedly selected by Borodin under the influence of Zinin, whose works, as is well known, were devoted to aromatic compounds of the benzene series.

Borodin's investigation of benzidine derivatives helped to a considerable extent in gaining an understanding of the complex problem involved in the mechanism of the benzidine,[31] diphenyline,[32] and semidine rearrangements. These investigations were of great significance both from the theoretical and from the practical points of view, since benzidine and its derivatives are widely used in the chemical industry, mainly for the production of some important organic dyes (especially substantives).[33]

Borodin wrote a detailed account of his scientific work during this period to the Head of the Medical-Surgical Academy, Pëtr Aleksandrovich Dubovitskiĭ[34].

In April 1860 Borodin went on a trip to Wiesloch[35], where he inspected some zinc mines. During the summer holidays, when the laboratory was being cleaned, he studied Italian, read Russian journals and books, and traveled with companions in the picturesque environs of Heidelberg. Once a week he took part in the performance of a quartet or quintet. At this same time (spring 1860) he composed a string sextet[36], which was performed for the first time in Heidelberg. Borodin wrote to his mother,

> I finally passed for a musician here ... Even the aristocratic circle took it into their head all of a sudden to admit me. Well, of course, I no longer go there as they are really painfully stupid[37].

In his Avtobiograficheskie zapiski (Autobiographical Notes) I. M. Sechenov wrote:

> I remember, for example, that in Mendeleev's apartment "Obryv" (The Precipice) by [Ivan Aleksandrovich] Goncharov, published at this time, was read aloud, that the audience eagerly listened to it and that, pressed by intellectual

hunger, it seemed to be the peak of perfection. I remember that Borodin, having a piano in his apartment, sometimes entertained the audience with music, carefully concealing the fact that he was a serious musician by never playing anything serious, but only, according to the wishes of the listeners, some songs or beloved arias from Italian operas. Knowing that I was passionately fond of "The Barber of Seville,"[38] he treated me to all the main arias of this opera and in general amazed all of us by the fact that he was able to play everything that we requested without music, from memory . . .[39].

Recalling later the days in Heidelberg, one of the contemporaries who lived there together with Borodin, wrote:

The soul rests when we think about Aleksandr Porfir'evich Borodin, the noted Russian chemist and composer. He was a delightful, seductive person, generously gifted by nature. Whatever he undertook, everything of his came out right. He mastered foreign languages with amazing rapidity: having lived in Paris, he grew to speak French like a street *gamin*; having stayed in Italy, he studied all its slang. And his salutation was such a simple and comradely one, as if he were the most commonplace person[40].

In the beginning of August of that same year, accompanying Professor M. Ya. Kittary's[41] wife to Bonn, Borodin passed through Holland and saw the

Fig. 5. Group of members of the Heidelberg Circle (from left to right): Zhitinskiĭ, Borodin, Mendeleev, Olevinskiĭ

sights of Rotterdam. Upon his return to Heidelberg he met Zinin, who had returned from Paris and was living at this time in Heidelberg[42]. "We are now inseparable from him," reports Borodin, "and are waiting for good weather in order to start off for Switzerland and Italy"[43]. Borodin, learning from Zinin that Hofmann[44] was working in London on those same problems that also interested him, hurried to finish his work, Issledovanii nekotorykh proizvodnykh benzidina (On Investigations of Some Derivatives of Benzidine) [*B6*].

On August 24, 1860 Borodin, Zinin, and Mendeleev went to Italy. They went from there to Karlsruhe[45] where the International Congress of Chemists was to take place from September 3 to September 6, 1860. Scholars from almost every country (127 persons in all) attended the Congress. Zinin, Mendeleev, Borodin, Savich, [Leon Nikolaevich] Shishkov, Lesinskiĭ, and [Yakub] Natanson represented Russia.

Footnotes to Chapter 4

[1] See Appendix I. — A.

[2] Pis'mo-nakaz Zinina doktoru Borodinu (Letter of Instructions by Zinin to Doctor Borodin), November 13, 1859, Sheet 21, Military-Historical Archives of the USSR in Leningrad. It is published here for the first time. See Appendix II. — A.

[3] For example, Mendeleev was able to equip his own laboratory, although small, with his mission money. — A.

[4] Aleksandr Mikhaĭlovich Butlerov (1828–1886) was a great Russian chemist. He was the creator of the theory of the "chemical structure of organic substances" and was a student of N. N. Zinin's. From 1869 he was a professor at St. Petersburg University and from 1874 he was ordinary academician of the Academy of Sciences. — A. [See G. V. Bykov: Aleksandr Mikhaĭlovich Butlerov. In: DSB, 1970, Vol. 2, pp. 620–625 and Henry M. Leicester: Alexander Mikhaĭlovich Butlerov. Journal of Chemical Education. *17*: 203–209 (1940). — T.]

[5] August Kekulé (1829–1896) was a German chemist and a professor at Bonn University. — A. [See Richard Anschütz: August Kekulé, Vol. I, Leben und Wirken; Vol. II, Abhandlungen, Berichte, Kritiken, Artikel, Reden. Verlag Chemie, Berlin, 1929 and Jean Gillis: (Friedrich) August Kekulé von Stradonitz. In: DSB, 1973, Vol. 7, pp. 279–283. — T.]

[6] Leon Nikolaevich Shishkov (1830–1909) was a chemist. In 1851 he graduated from the Artillery Academy. From 1860 to 1865 he was Professor of Chemistry at the Artillery Academy, where he organized a first-class chemical laboratory. He was a close friend of Mendeleev's and was well known as an outstanding researcher in the field of explosive substances. — A. [See GSE, 1982, Vol. 29, p. 606. — T.]

[7] M[ikhail] N[ikolaevich] Mladentsev and V[yacheslav] E[vgen'yevich] Tishchenko, [Dmitriĭ Ivanovich Mendeleev, ego zhizn' i deiatel'nost'] (D. I. Mendeleev, [His Life and Work], hereafter abbreviated as Mendeleev), Vol. 1, Moscow, 1938, p. 161. — A.

[8] Borodin: Letters, I, p. 34. — A.

[9] Emil Erlenmeyer (1825–1909) was a German organic chemist. — A. [See Albert B. Costa: Richard August Carl Emil Erlenmeyer. In: DSB, 1971, Vol. 4, pp. 397–400. — T.]

[10] Karpfenstrasse 6. — T.

[11] M. N. Mladentsev and V. E. Tishchenko: Mendeleev, p. 172. — A.

[12] For additional details see David Lloyd-Jones: Borodin in Heidelberg. The Musical Quarterly. *46* (4): 500–508 (1960). — T.

[13] Valerian Savich (?–1862) was a chemist and a close friend of Mendeleev's and Borodin's. — A.

[14] Olevinskiĭ (?–1860) was a medical man by education, working as a junior intern at the Nikolaevskiĭ Naval Hospital. In the 1860's he was abroad, where he studied chemistry. He was a friend of Mendeleev's and Borodin's. — A.

[15] Stepan Vasil'evich Eshevskiĭ (1829–1865) was a professor of history — A. [who specialized in the history of the late Roman Empire and the early Middle Ages. See GSE, 1975, Vol. 9, pp. 136–137. — T.]

16 Tat'iana Petrovna Passek (1810–1889) was a writer. She was the author of a volume of recollections: Iz dal'nikh let (From Distant Years) and a relative of A[leksandr] I[vanovich] Herzen's. — A.

17 Marko-Vovchok (1834–1907) was the literary pen-name of Maria Aleksandrovna Velinskaia, a well-known Ukrainian writer. — A.

18 Eduard Andreevich Yunge (1833–1898) was an oculist, professor at the Medical-Surgical Academy, and a friend of Borodin's. — A.

19 Pëtr Petrovich Alekseev (1840–1891) was a famous Russian chemist and a professor at Kiev University. He wrote Uchebnik organicheskoĭ khimii (Textbook of Organic Chemistry) and was a friend of Borodin's. — A. [He discovered a method for preparing azobenzene, which is of great importance in industrial and laboratory work with all classes of nitrogen compounds. See GSE, 1973, Vol. 1, p. 229. — T.]

20 Aleksandr Onufrievich Kovalevskiĭ (1840–1901) was a noted Russian embryologist and academician. — A. [See Mark B. Adams: Aleksandr Onufrievich Kovalevsky. In: DSB, 1973, Vol. 7, pp. 474–477. — T.]

21 Lev Semenovich Tsenkovskiĭ (1822–1887) was a famous Russian naturalist. He was the founder of scientific bacteriology and the author of outstanding works on morphology. — A. [See GSE, n. d., Vol. 28, p. 688. — T.]

22 Andreĭ Sergeevich Famintsyn (1835–1918) was a botanist and academician. — A. [See GSE, 1979, Vol. 23, pp. 196, 651; 1981, Vol. 27, pp. 95, 573. — T.]

23 Konon Ivanovich Lisenko (1837–1903) was a professor of chemistry at the [St. Petersburg] Gornyĭ Institut (Mining Institute). — A. [See GSE, 1977, Vol. 14, p. 538. — T.]

24 Borodin: Letters, I, p. 36. — A.

25 Letter from A. P. Borodin to P. P. Alekseev, May 24, 1861, published here for the first time. — A.

26 M. N. Mladentsev and V. E. Tishchenko: Mendeleev, pp. 166–167. — A.

27 This periodical was published in St. Petersburg from 1836 to 1866. — T.

28 The first uncensored Russian periodical, published from 1857 to 1867 by Alexander Herzen (in Russian, Aleksandr Ivanovich Gertsen), journalist and political thinker, who originated the theory of a unique Russian path to socialism, or peasant populism. — T.

29 T. V. Volkova: Letters of A. P. Borodin [hereafter abbreviated as Letters]. Uspekhi khimii (Advances in Chemistry).*9* (9): 1066 (1940). — A.

30 Borodin: Letters, I, pp. 36–37. — A.

31 Benzidine or *p,p'*-diaminobiphenyl, $(H_2N \cdot C_6H_4)_2$, the most important derivative of biphenyl, is made by the benzidine rearrangement of hydrazobenzene:

$$C_6H_5NH \cdot HNC_6H_5 + 2\,HCl \xrightarrow{\text{Heat}} [\,H_3NC_6H_4 \cdot C_6H_4 \cdot NH_3\,]\,Cl_2$$

Benzidine hydrochloride

If one or both *para* positions in the hydrazo compound are blocked, an amino diphenylamine (semidine) is formed instead of a biphenyl derivative.

The mechanism of this well-investigated rearrangement has been difficult to explain because the rearrangement is entirely intramolecular and various products are formed under different conditions. No simple proposed mechanism explains all the facts, and several different mechanisms may be involved (See P. A. S. Smith: Open Chain Nitrogen Compounds. W. A. Benjamin, New York, 1966, Vol. 2, pp. 123–126). — T. The translators are indebted to Prof. Henry J. Shine of Texas Tech University, who brought to our attention the fact that Borodin [*B 5*, *B 6*] did not investigate the mechanism of these rearrangements. According to Prof. Shine, in recent years, these rearrangements have been shown to be sigmatropic ones, whose mechanisms are governed by the principles of conservation of orbital symmetry (see H. J. Shine *et al.*, J. Am. Chem. Soc., *104*, 2501–2509 (1982); *106*, 7077–7082 (1984); *107*, 3218–3223, 6674–6678 (1985); *108*, 1000–1006 (1986)).

32 Biphenyline or 2,4'-diaminobiphenyl, 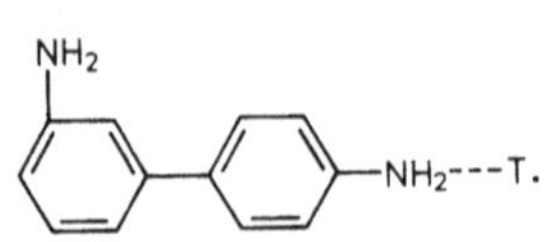

[33] Dyes attaching directly to the material without the aid of a mordant, as opposed to adjective dyes, which require a mordant to render them permanent. — T.

[34] See Appendix III. — A.

[35] A small town about 6 miles (ca. 10 km) south of Heidelberg. — T.

[36] Sextet for Two Violins, Two Violas, and Two Violoncelli in D Minor. — T.

[37] Borodin: Letters, I, p. 38. — A.

[38] Il Barbiere di Siviglia, the opera by Gioacchino Rossini (1792–1868). — T.

[39] Sechenov: Autobiographical Notes, p. 104. — A.

[40] A[leksandr] V[asil'evich] Romanovich-Slavatinskiĭ: Moia zhizn'. Vestnik Evropy (My Life. Notes of Europe), 1903, April, Chapter 8, p. 539. — A.

[41] Modest Yakovlevich Kittary (1825–1880) was a famous Russian technologist. From 1857 he occupied the Chair of Technology in Moscow University and was the director of the Moscow Practical Academy of Commercial Sciences. — A.

[42] According to the Commission of the Medical-Surgical Academy, Zinin traveled during the summer of 1860 in order to inspect the buildings and equipment of foreign chemical laboratories. — A.

[43] Borodin: Letters, I, p. 52. — A.

[44] August [W. von] Hofmann (1818–1892) was a German chemist who for several years after 1845 lived and worked in England. — A. [See W. H. Brock: August Wilhelm von Hofmann. In: DSB, 1972, Vol. 6, pp. 461–464. — T.]

[45] A city about 26 miles (ca. 42 km) southwest of Heidelberg. The Karlsruhe Congress, organized by Carl Weltzien (1813–1870), who was born in St. Petersburg, has generally been considered the first international scientific congress. However, this has been disputed by Maurice Crosland [The Congress on Definitive Metric Standards, 1798–1799: The First International Scientific Conference?. Isis. *60*: 226–231 (1969)]. — T.

Chapter 5

Travel in Europe

The International Congress of Chemists in Karlsruhe was an extremely important event in the history of chemistry. Until the middle of the last century complete disorder reigned in the explanations and interpretations of basic chemical ideas (atom, molecule, equivalent). Never-ending arguments among chemists concerning these questions were stirred up by the principal difference between the spokesmen of the two chemical theories: the old electrochemical theory, established during the 1820's by Jöns Jacob Berzelius, and the new unitary theory, proposed by Charles Frédéric Gerhardt and Auguste Laurent.

During the first half of the last century chemists imagined that each chemical compound consisted of two components. The noted Russian scholar Butlerov wrote,

> Chemistry was for the most part the chemistry of oxygen compounds. These compounds were considered as consisting of oxygen and a radical generally understood as the component part combined with oxygen[1].

In 1820 Berzelius stated and developed his electrochemical theory of the structure of chemical compounds[2]. According to his theory, each complex body consists of two electrically opposite parts, for example, potash, K_2CO_3, from positively charged K_2O and negative CO_2; sulfuric acid, H_2SO_4, from positive H_2O and negative SO_3.

Twenty years after the appearance of Berzelius' theory, the French chemist Jean Baptiste André Dumas published the results of his research. Dumas obtained a new substance by the chlorination of acetic acid, the study of which showed that the product of the reaction consisted of CCl_3COOH (trichloroacetic acid), an acid similar to but differing slightly from acetic acid. We must note that the famous Russian chemist, T. E. Lowitz, had obtained trichloroacetic acid for the first time by the same method as early as the end of the eighteenth century. By this method, negatively charged chlorine replaced positively charged hydrogen, but, according to the electrochemical theory, the CCl_3 group must be considered as positively charged. Guided by this research, Dumas developed the doctrine of metalepsy (substitution) and laid the foundations for the co-called type theory of chemical compounds. The French chemist Gerhardt developed this doctrine extensively. He proposed his unitary theory, according to which every chemical compound was considered, in contrast to Berzelius' theory, as an entity, the atoms of which were subject to the laws of substitution.

These views underwent fierce attacks from the chemists of Berzelius' school and from Berzelius himself, whose scientific authority hindered the recognition of the new theory. Great disagreements arose among chemists. The concepts of equivalent weight, molecular weight, and atomic weight were used in textbooks and articles arbitrarily, often being confused. This, or course, greatly hindered the development of chemistry.

The Congress had to settle the following questions:

1. By means of an exchange of ideas and discussion of the individual principal questions to decide which of the contemporary theories deserves preference.
2. To reach an agreement or at least to make preparations to express the same ideas in the same form, both in words and in writing; for example:
   a) to establish what words are necessary ... for the definition of concepts, for example: equivalent, atom, molecule, atomic, basic (basicity), atomicity,[3] two volume or four volume, etc.;
   b) by what symbol to designate atoms and equivalents of elements ...
   c) agreement concerning the manner of writing rational formulae ...
   d) preparation of a uniform and rational nomenclature[4].

At the Congress "Gerhardt's followers, of whom the Italian professor Cannizzaro[5] was the leader, eagerly showed the consequences of Avogadro's Law ..." Mendeleev wrote,

> I vividly remember the effect of his speeches, in which there were no compromises, but the truth itself, taken from the outcome of the concepts of Avogadro[6], Gerhardt, and Regnault[7], at that time still far from being accepted, was heard[8].

Later on Mendeleev wrote:

> Not a few years have passed since Cannizzaro's ideas were found to be the only ones able to endure criticism and to give a conception of atoms as the least quantity of elements, which entered into particles of their compounds[9].

Mendeleev wrote a long letter about the Congress to his teacher, Voskresenskiĭ[10], in which he gave a detailed analysis of the work of the Congress and the decisions adopted by it[11].

The Congress' decision and the recognition of Avogadro and Gerhardt's Law by the majority of the chemists, which resulted in the decision to use atomic weights in place of equivalents in chemical calculations, played a large role in the subsequent development of chemistry. Zinin's brilliant lectures at the Congress sessions and Mendeleev and Shishkov's active participation as members of the committee for drawing up the decisions of the Congress drew the attention of the members of the Congress to the Russian delegates and raised the importance of Russian chemical science in their eyes.

The Congress made a great impression on the 27-year-old Borodin. Contact with outstanding scientists and the triumph of the unitary system, which he had already earlier supported, inspired him to carry out subsequent, extensive investigations in the field of chemistry.

After the Congress Zinin returned to Russia and Mendeleev and Borodin to Heidelberg. In October Borodin decided to go to Italy with Mendeleev. In a letter to his mother Borodin relates:

> To tell the truth, I left Heidelberg, where I spent almost an entire year so peacefully and well, with a little sadness. It is true that except for Erlenmeyer I was acquainted with practically no Germans. However, our Russian circle lived here in a truly friendly manner, amicably obliging one another mutually with what one could. I doubt whether you will find such a close and friendly circle anywhere else. However, it seems that in Paris we all will assemble anew. Having completely packed, I took final leave of my good friends and started the journey[12].

Recalling this trip, Mendeleev stated:

> We started on our way with very little baggage, with only one miniature traveling bag for the two of us. We wore only blouses in order to resemble artists, which was very advantageous in Italy. During the spring of 1860 we visited Venice, Verona, and Milan; in the autumn of that same year Genoa and Rome, after which Borodin went to Paris for a short time. On the first trip a funny incident happened to us on the railroad. Near Verona the Austrian[13] police began to search our car and to look for someone: we were informed that a political criminal was on this trip, an Italian who had just escaped from prison. They immediately took Borodin for this criminal because of the southern constitution of his physiognomy. They went through all of our meager baggage, questioned us, and wanted to make an arrest but soon were convinced that we really were Russian students and left us in peace. Such was our amazement when, leaving the Austrian border and entering Sardinia, we were made the object of a whole celebration; everyone in the car embraced and kissed us, shouting "Viva" and sang at the tops of their voices. The reason for this was that the political criminal was sitting in our car all the time, only they didn't notice him and he happily escaped the Austrian claws.
>
> After a stifling, cloistered life in Heidelberg, we fully enjoyed ourselves in Italy. We ran along the streets the whole day, looking into churches, museums, but most of all we loved the small folk theaters, which delighted us by their liveliness, gaiety, and the unbridled comedy typical of truly native performances[14].

The Italian people, on whose lips was already the name of Garibaldi[15], drew Borodin to their own selfless struggle for the unity and freedom of their native land. Mendeleev wrote from Italy:

> I lived among a lively, refined, but also poor people, suffering under the yoke of the Pope and not being able to break this yoke because thirty thousand French soldiers are ready to defend it, prepared to raid the city from Fort St. Angel[16, 17].

In Rome Borodin and Mendeleev viewed the art monuments. They parted in Livorno[18] at the end of November: Borodin went to Paris, Mendeleev to Heidelberg.

Shortly afterward, a large group of Russians met in Paris. In addition to the members of the Heidelberg Circle there were Nikolaĭ Nikolaevich Beketov[19], Savich, Pëtr Petrovich Alekseev, Dmitriĭ Nikolaevich Abashev[20], and others. In the French capital well-made chemical apparatus and reagents were easily obtained. But Paris attracted the Russian youth not only by its laboratories and scientists. The freedom-loving French people, having had to struggle for their liberation several times, could not help but arouse the admiration of the progressively minded young Russians, who craved a radical rebuilding of the social-political system for themselves at home.

Recalling Paris, Mendeleev wrote:

> The people, that is, the workers of Paris, are a new breed for me, interesting in every respect. These people, compelled to tremble by the kings, are striking: they are honest-minded, even refined; they read a lot, and talk about everything; they live day by day. This is a veritable people of life; you can understand how anyone would be terrified if crowds of such people were to rouse themselves. This class is quite different from the bourgeoisie, from the tradesmen, that sweet, frivolous class . . . not that class to whom the history of France belongs[21].

Borodin remained in Paris for several months. During this time he profited by becoming acquainted with a new method for studying the structure of organic compounds (polarimetry), he listened to Regnault's and Bernard's[22] lectures and apparently worked in one of the Parisian chemical laboratories[23].

In 1860 he published an article Issledovanie o deĭstvii broma na serebrianye soli kislot: uksusnoĭ, maslianoĭ i valerianovoĭ (Investigation of the Action of Bromine on Silver Salts of Acids: Acetic, Butyric, and Valeric)[24] [*B 7*]. In this work new compounds were described: the bromo derivatives of carboxylic acids and also other exceptionally interesting substances, namely, the mixed anhydrides of hypobromous acid and the lower fatty acids [acyl hypobromites, RCO_2Br. — T.].

Borodin's work on halogen derivatives of acids remained unfinished. At this time Schützenberger's[25] article appeared, in which compounds of hypochlorous acid were described, similar to those obtained by Borodin. Therefore he decided to stop his investigations in this field temporarily.

During the spring of 1861 Borodin again undertook a trip to Italy. He visited Naples and saw the production of boric acid in Volterra.[26] "For our part," wrote Borodin to Alekseev after his return home, "I saw 'Lagoni' [Marshes] and the manufacture of boric acid, and I brought back a collection of lava from Vesuvius for the Academy"[27].

Borodin returned to Heidelberg in May. Life in Heidelberg had outwardly changed little, but the Heidelberg Circle, which had been connected with so many pleasant memories, had already collapsed. Mendeleev had gone to St. Petersburg in February, V. I. Olevinskiĭ had died. Many of Borodin's friends had departed.

Life in Heidelberg did not favor the development of Borodin's musical talent. No serious musicians were there, and Borodin felt lonely in this respect. In addition to this, his friends in chemistry often spoke ironically about his enthusiasm for music. It is interesting to note that nowhere in his letters to chemists of this period does Borodin mention music. Here is, as Borodin himself described, the cultural condition in Heidelberg:

> The city has entertainment: theaters and concerts of the Symphonic Society. I managed to attend one of these concerts. The music was very pretty. But the theater is simply the devil knows what! Besides two or three characters, the rest are not fit at all. And the play "Der Verschwender"[28] [The Spendthrift] which I saw, was truly astonishing. It is difficult to think of something more senseless. But the Germans sat enraptured[29].

During the spring of 1861 Borodin became acquainted with Ekaterina Sergeevna Protopopova, a talented pianist, who arrived in Heidelberg for a cure[30].

Ekaterina Sergeevna Protopopova's arrival in Heidelberg and his meeting with her were great events for Borodin. The performance of Protopopova, who possessed a perfect piano technique[31] and who knew the works of composers yet little known to Borodin, opened a new world of sound to him. Protopopova's penetrating performance of the songs of Chopin, Schumann, and Liszt, her favorite composers, captivated and charmed him.

During the summer Borodin and Protopopova often traveled in the mountains, where they visited many old castles. Later, recalling her first acquaintance with Borodin in Heidelberg, Protopopova wrote:

> We were often together. His day was organized so that from 5:00 in the morning until 5:00 in the evening he was in the chemical laboratory, from 5:00 to 8:00 we took trips in the mountains, and from 8:00 or 9:00 in the evening until 12:00 we had music in the hall of Hofmann's boarding house[32].

On holidays the young people went to Mannheim[33], where there was an opera house.

Soon, at the insistence of her physicians, Protopopova had to go south:

> With the approach of autumn and cold weather it again became worse for me, after my summer respite. A. P. [Borodin] left the Heidelberg laboratory for several days in order to accompany and settle me in Pisa. An Italian October met us there: heat, perfect summer, the days flew by. Some panic to be left alone came over me. Aleksandr [Borodin] packed his things, and at the last minute he went on an official visit to two famous Pisa chemists, Luca[34] and Tassinari[35]. I was left alone. It is impossible to describe how painful it was for me. Suddenly I heard Aleksandr's voice: "Katia,[36] imagine what has happened! I am not going to Heidelberg. I will remain here with you the whole time. Luca and Tassinari received me in a most amiable manner. Their laboratory is magnificent, bright, and comfortable. They offered it to me at my complete disposal. And you know how good this will be: the fluorine compounds, which I have now begun to study require experiments to be carried out in the open. In Heidelberg it is too cold for this, but here I can study them all winter"[37].

For the first time Borodin succeeded in obtaining benzoyl fluoride by the action of potassium hydrogenfluoride, KHF_2, on benzoyl chloride in Luca and Tassinari's laboratory[38] [*B10*]. This investigation is remarkable, first of all because Borodin was [one of] the first chemists to prepare an organic fluorine compound,[39] second, because he worked out a method for fluorinating organic compounds, and finally, because he showed that fluorine corresponds completely to the other halogens in carbon compounds.

Borodin's work on the fluorination of organic compounds involved enormous experimental difficulties. As is well known, fluorine compounds (both inorganic and organic) are highly reactive chemical substances requiring special chemical vessels and expensive laboratory equipment. The presence of suitable conditions in the chemical laboratory of Pisa University allowed Borodin to carry out this extremely difficult work.

Protopopova recalls,

> Chemistry did not prevent Aleksandr from also giving some time to music. For example, he played the 'cello in the orchestra of the Pisa theater . . . In Pisa we were acquainted with the director of the local music school, Signore Menocchi. He was an obliging man but not an exceptional musician. I remember that once, in his presence, Aleksandr jotted down a fugue in no more than a few hours. You should have seen the amazement of the Signore Professore. From then on he began to look at A. P. as a musical marvel[40].

During the time of his stay in Pisa, Borodin composed a tarantella for piano for four hands[41] and a quintet for strings and piano.[42] These works were completed during the summer of 1862, a few months before his departure for Russia.

The nearer the date of the termination of his mission approached, the more Borodin felt himself to be "a visitor" abroad. He wrote to Mendeleev:

> I begin to think about Piter [St. Petersburg], about arranging a laboratory. I need to acquire a home laboratory . . . For this goal I must acquire some pieces of equipment. I have already ordered from Shapiui[43] two platinum retorts, . . . but a microscope is necessary. I ordered an air-pump with a glass cylinder and a cast iron mounting and a hand pump (without racks and pinions) with two large bells (with stopcocks above) and with a manometer. As you see, I have been buying [equipment]. I do not think they will grant me an extension although I would greatly like one. However, at present I do not really know that I want it. What will be later, I do not know[44].

Borodin's situation was also characteristic of other young Russian scientists who were abroad. Conflicting attitudes arose for various reasons. On the one hand, finding themselves far from Russia, they felt that they were free men, not experiencing the severe atmosphere of the Czarist bureaucratic regime. On the other hand, they were drawn by their native land. Often they were ready to give up all their activity in order to return to their native surroundings. "I was sometimes so terrified that I would have given up everything and flown to Russia", wrote Mendeleev[45].

Other scientists of the Heidelberg Circle who remained in Germany also experienced this attitude, yearning for their native land and suffering from the unfavorable news from Russia which reached them in letters from friends. A friend of Mendeleev's, A. Baryshnikov, wrote: "I can tell you nothing new about Russia. Everything here is still the peasant question, the same shares, bonds, abuse by bribe-takers, officials. Few [people] are working"[46]. For an obvious reason Baryshnikov was not able to write about the student agitations in the universities or about the peasant "mutinies". But Borodin doubtless knew about them from comrades and in part from Sechenov, who returned to Russia earlier than Borodin and Mendeleev and with whom Borodin was in lively correspondence.

Remaining abroad for 3 years, Borodin carried out eight original chemical works, among them the outstanding investigations, Poluchenie galoidzameshchennykh carbonovykh kislot deĭstviem broma na serebrianye soli kislot: uksusnoĭ, maslianoĭ i valerianovoĭ (The Production of Halogen-Substituted Carboxylic Acids by the Action of Bromine on the Silver Salts of the Acids: Acetic, Butyric and Valeric)[47] [*B7*]; O ftoristom benzoile i drugikh proizvodnykh ftora (On Benzoyl Fluoride and Other Derivatives of Fluorine) [*B10*]; O deĭstvii slozhnykh efirov na tsinketil (On the Action of Esters on Zinc Ethyl) [*B8*], and others. These researches provided the author with well-deserved fame in Russia as well as abroad[48].

Footnotes to Chapter 5

[1] N. D. Zelinskiĭ: A. M. Butlerov i sovremennost' (A. M. Butlerov and the Present). Oktiabr' (October). *8*: 172 (1948). — A.

[2] Berzelius gave a sketch of his dualistic electrochemical theory in 1811, but he first developed it fully in 1818 in the first edition of Volume 3 of his Lärbok i Kemien. — T.

[3] Atomicity was the term used by Kekulé and others to designate what we today call valency or valence. — T.

[4] M. N. Mladentsev and V. E. Tishchenko: Mendeleev, p. 245. — A.

[5] Stanislao Cannizzaro (1826–1910) was a famous Italian chemist. — A. [See Henry M. Leicester: Stanislao Cannizzaro. In: DSB, 1971, Vol. 3, pp. 45–49. — T.] In 1858 he wrote the book: Sunto di un corso di filosofia chimica, which played a large role in the development of the atomic theory. — A. [Originally published in Il Nuovo Cimento. *7*: 321–366 (1858); for an English translation see: Sketch of a Course of Chemical Philosophy. Alembic Club Reprint No. 18, E. & S. Livingstone, Ltd., Edinburgh, 1960. — T.]

[6] Amedeo Avogadro (1776–1856) was a famous Italian chemist and physicist. — A. [See M. P. Crosland: Amedeo Avogadro. In: DSB, 1970, Vol. 1, pp. 343–350 and Mario Morselli: Amedeo Avogadro: A Scientific Biography. D. Reidel Publishing Co., Dordrecht, Boston, Lancaster, 1984. — T.]

[7] Victor Regnault (1810–1878) was a French physicist and chemist. — A. [See Robert Fox: Henri Victor Regnault. In: DSB, 1975, Vol. 11, pp. 352–354. — T.]

[8] M. N. Mladentsev and V. E. Tishchenko: Mendeleev, p. 259. — A.

[9] *Ibid.* — A.

[10] Aleksandr Abramovich Voskresenskiĭ (1809–1880) was an outstanding Russian chemist. He was Professor of Chemistry at St. Petersburg's Main Pedagogical Institute. Among his students were D. I. Mendeleev, N. N. Beketov, N. N. Sokolov, N. A. Menshutkin, P. P. Alekseev, and others. — A. [See Charlene Steinberg: Alexandr Abramovich Voskresenskiĭ, Grandfather of Russian Chemists. Journal of Chemical Education. *42*: 675–677 (1965). — T.]

[11] See Appendix IV. — A.

12 Borodin: Letters, I, pp. 53–54. — A.
13 Before its unification, much of Italy was under Austrian rule. — T.
14 V. V. Stassov: Borodin, p. 148. — A.
15 Giuseppe Garibaldi (1807–1882) was a noted Italian patriot who devoted his entire life to the struggle for the unity and freedom of Italy. — A. [See John Parris: The Lion of Caprera. D. McKay Co., New York, 1962; Dennis Mack Smith: Garibaldi: A Great Life in Brief. Prentice-Hall, Englewood Cliffs, N. J., 1969; and David Glass Larg: Giuseppe Garibaldi. Kennikat Press, Port Washington, N. Y., 1974. — T.]
16 The Castel Sant' Angelo (Hadrian's Tomb) on the right bank of the Tiber River in Rome was built in A. D. 135–139 and converted into a fortress in the fifth century. It is best known in music as the location of Act III of Giacomo Puccini's opera *Tosca*. — T.
17 M. N. Mladentsev and V. E. Tishchenko: Mendeleev, p. 198. — A.
18 Conventionally called Leghorn in English. Capital of Livorno province, Toscana (Tuscany) region, central Italy. — T.
19 Nikolaĭ Nikolaevich Beketov (1827–1911) was an outstanding Russian physical chemist. He was Professor of Chemistry at Kharkov University until 1887; from 1887 he was an academician. — A. [See G. V. Bykov: Nikolaĭ Nikolaevich Beketov. In: DSB, 1970, Vol. 1, p. 579. — T.]
20 Dmitriĭ Nikolaevich Abashev (1829–1880) was an agricultural chemist who was a professor at Kiev University. — A.
21 M. N. Mladentsev and V. E. Tishchenko: Mendeleev, pp. 186–187. — A.
22 Claude Bernard (1813–1878) was a French physiologist. — A. [See M. D. Grmek: Claude Bernard. In: DSB, 1970, Vol. 2, pp. 24–34. — T.]
23 See Appendices V [and VI]. — A.
24 CH_3COOH, $CH_3(CH_2)_2COOH$, and $CH_3(CH_2)_3COOH$, respectively. — T.
25 Paul Schützenberger (1827–1897) was a French chemist. — A. [He was professor at the École Supérieure de Chimie in Mulhouse and later at the Collège de France in Paris. — T.]
26 Town and episcopal see, Pisa province, Tuscany region, northwest of Siena in central Italy. — T.
27 Letter of A. P. Borodin to P. P. Alekseev, May 21, 1861, published here for the first time. — A.
28 This play, The Spendthrift, is considered one of the best works of the Viennese playwright Ferdinand Raimund (1790–1836), who refined the types of the commedia dell'arte and turned them into native characters. His plays are still performed today. — T.
29 Borodin: Letters, I, p. 36. — A.
30 The daughter of the staff physician of the Golitsyn Hospital, Sergeĭ Stepanovich Protopopov, and his wife, Ekaterina Alekseevna (née Konstantinova), daughter of the hospital supervisor, Ekaterina Sergeevna Protopopova was born in the Golitsyn Hospital in Moscow on January 3, 1832 and died in the village of Ramenskoe, near Moscow, on June 28, 1887, slightly more than four months after Borodin's death. She traveled to Heidelberg to undergo climatic treatment for incipient tuberculosis, using money raised by a piano recital that she gave in Moscow on April 27, 1861. Throughout her marriage with Borodin she was plagued with severe asthma attacks, hypochondria, and insomnia, and she often lived in Moscow, while he resided in St. Petersburg. Because of her ill health she did not go to sleep until 3 or 4 A.M., and Borodin could not get to sleep before that time. She arose about 3 or 4 P.M. the following day. As a result of this peculiar way of life Borodin suffered from lack of sleep, and this unusual routine may have hastened the heart disease which led to his premature death. In Heidelberg she stayed at Professor Hofmann's pension at Bergheimerstrasse 14, where Borodin soon moved from his lodgings with his friend N. M. Yakubovich on the Schulgasse. — T.
31 She was endowed with absolute pitch (the ability to sing or recognize the pitch of a tone by ear). According to Protopopova's memoirs, during a concert in Baden-Baden Borodin realized that she possessed this ability, and "it was that very evening that we both knew for certain, though we never admitted it to each other, that we were in love". — T.
32 V. V. Stassov: Borodin, p. 18. — A.
33 City on the right bank of the Rhine River opposite Ludwigshafen at the mouth of the canalized Neckar River about 11 miles (ca. 18 km) northwest of Heidelberg. — T.
34 Sebastiano de Luca (1820–1880) was an Italian chemist. — A.
35 Paolo Tassinari (1829–1909) was an Italian chemist. — A. [For biographical information see Ber. *42*: 2184–2185 (1909). — *T.*]
36 Katia is the diminutive for Ekaterina (Catherine). — T.

37 V. V. Stassov: Borodin, p. 20. — A.

38 $C_6H_5COCl + KHF_2 \rightarrow C_6H_5COF + HF + KCl$. Also, $C_6H_5COCl + KF \rightarrow C_6H_5COF + KCl$. — T.

39 The first organic fluorine compound, methyl fluoride, was synthesized by Jean Baptiste André Dumas and Eugène Melchior Péligot [Ann. Chim. Phys. [2] *61*: 193 (1836)]:

$$CH_3SO_3OK + KF \rightarrow CH_3F + K_2SO_4$$

The next member of the alkyl fluoride series, ethyl fluoride, was prepared by Edmond Frémy two decades later [Ann. Chim. Phys. [3] *46*: 1748 (1856); [3] *47*: 5 (1856)]:

$$C_2H_5SO_3OK + KHF_2 \rightarrow C_2H_5F + HF + K_2SO_4$$

Although Borodin prepared benzoyl fluoride by a general method applicable to acyl fluorides in 1862, Wilhelm Foerst [Ullmanns Encyclopädie der technischen Chemie, 3rd rev. ed., Urban & Schwarzenberg, Munich, 1956, Vol. 7, p. 607] credits the discovery of the first aromatic fluorine compound to W. Lenz in 1877 [Über Jodbenzol-Sulfonsäure. Berichte. *10*: 1135–1137 (1877), p. 1137] – 15 years later than Borodin's discovery.

In his study of the metathesis between benzoic acid and potassium fluoride, Borodin [Il Nuovo Cimento. *15*: 305–314 (1862); Compt. rend. *55*: 553–556 (1862); Chem. News. *6*: 267—268 (1862)] [*B 10*] observed the formation of the acid double fluoride:

$$C_6H_5COOH + 2\,KF \rightarrow C_6H_5COOK + KHF_2$$

and concluded that hydrogen fluoride must be a dibasic (diprotic) acid, a view apparently supported by John William Mallet's determination of its molecular weight as 39.32 [American Chemical Journal. *3*: 189–197 (1881)]. The formation of hydrogenfluoride salts, e.g., KHF_2, was later accounted for more satisfactorily by the concept of hydrogen bonding. — T.

40 V. V. Stassov: Borodin, pp. 19–20. — A.

41 Tarantella in D Major for piano duet. See p. 129. — T.

42 Piano Quintet in C Minor. See p. 128. — T.

43 A manufacturer of apparatus. — T.

44 T. V. Volkova: Letters, p. 1065. — A.

45 M. N. Mladentsev and V. E. Tishchenko: Mendeleev, p. 191. — A.

46 Ibid, p. 184. — A.

47 CH_3COOAg, $CH_3(CH_2)_2COOAg$, and $CH_3(CH_2)_3COOAg$, respectively. Both bromobutyric and bromovaleric acids decomposed on distillation, in contrast to their lower homolog, bromoacetic acid, which could not be obtained by this reaction [A. Borodin: Bull. Soc. Chim. France. 252–254 (1861); Z. Chem. und Pharm. *4*: 5–7 (1861); Liebigs Annalen. *119*: 121–123 (1861); J. prakt. Chem. *84*: 474–475 (1861)] [*B 7*]. Instead, Borodin obtained methyl bromide:

$$CH_3COOAg + Br_2 \rightarrow CH_3Br + CO_2\uparrow + AgBr\,.$$

This type of general reaction, which provides a convenient method for decreasing the length of a carbon chain by one carbon atom,

$$RCOOAg + Br_2 \rightarrow RBr + CO_2\uparrow + AgBr\,,$$

is usually credited to the Hunsdieckers [H. Hunsdiecker and C. Hunsdiecker: Über den Abbau der Salze aliphatischer Säuren durch Brom. Ber. *75*: 291–297 (1942); Robert G. Johnson and Robert K. Ingham: The Degradation of Carboxylic Acid Salts by Means of Halogen: The Hunsdiecker Reaction. Chemical Reviews. *56*: 219–269 (1956)]. Borodin's priority is completely overlooked except for one chemical paper championing his cause and proposing the name "Borodin reaction" (David R. Howton, Robert H. Davis, and Judd C. Nevenzel: Unsaturated Fatty Acids. III. Preparation of 1-C^{14}-Linoleic Acid. J. Am. Chem. Soc. *76*: 4970–4974 (1954)). — T.

48 See Appendix VI. — A.

Chapter 6

First Years of Professorial Activity

In September 1862 Borodin, together with Protopopova, returned to Russia. Immediately upon his arrival in St. Petersburg, he visited the laboratory of the Medical-Surgical Academy and called on his teacher Zinin.

One of Borodin's pupils, Alekseĭ Petrovich Dobroslavin, told of this meeting:

> I remember, as if it had just happened, that minute when we students of the second course saw him for the first time in the auditorium. A young man, handsome, in a summer coat, made his way by means of a slow, somewhat hesitant gait to the office of Professor Zinin. Shortly afterward, word spread throughout the auditorium that this was Borodin, who had just returned from abroad. All the students who were close to Zinin heard afterward about the return of his favorite pupil. Having such an expansive nature, Zinin's feelings toward all his students were very sincere, but toward Borodin they were even more sincere. He regarded Borodin as his spiritual son, and Borodin on his part regarded him as his second father. There was neither a scientific thought nor a method of procedure which the teacher and the pupil did not discuss.[1]

Immediately upon his return to his native land, Borodin took part in academic life. The construction of the building of the Natural Science History Institute of the

Fig. 6. View of the St. Petersburg Natural Science History Institute, Medical-Surgical Academy

Medical-Surgical Academy, which was conceived by Zinin, was taking place during these years. Borodin, as a principal co-worker of Zinin's, had to take upon himself a considerable amount of the anxiety connected with this construction. At the same time he continued his scientific work in the laboratory and prepared for the examinations for the rank of professor.

In October 1862 Borodin was elected to the temporary post of Assistant (Ad'iunkt) Professor of Chemistry by the Conference of the Academy, and on December 8th of the same year, after appropriate examination and the reading of a trial lecture on the theme O znachenii analiza meroiu pri meditsinskom issledovanii (On the Significance of Analysis by the Standards Affiliated with Medical Research), he was named Ad'iunkt-Professor (Assistant Professor). Academician Zinin, who in 1864 stopped reading courses on chemistry and remained in the academy only in the capacity of Direktor Khimicheskikh Rabot (Director of Chemical Works), entrusted him with the reading of the first course in inorganic chemistry and the second course in organic chemistry.

The young professor began his work vigorously. Borodin reported to Protopopova:

> I was extremely busy all this time. I wrote and calculated all week so that I came to look at ciphers with revulsion. This work consisted of an order of laboratory things from abroad, which will be fairly advantageous. Now each student will receive a full set of chemical cups and glasses, etc., and beyond that, more than 1000 rubles are left of the money with which many good instruments for the laboratory can be acquired[2].

Borodin translated several scientific books for the St. Petersburg publishing house of Wolf.[3] Furthermore, he read lectures on chemistry at the Lesnaia

Fig. 7. Aleksandr Porfir'evich Borodin (early 1860's)

akademiia (Forestry Academy). These works served for his subsidiary earnings. Borodin wrote to Protopopova:

> I am telling you in confidence of a pleasant hope, it seems they are going to increase my salary. That will be nice! Then I can cast the Forestry Academy and Wolf to the devil and enjoy life[4].

In 1863 Borodin began, as he expressed it, "a small chemical work", a significant investigation on the condensation products of valeraldehyde [C_4H_9CHO], enanthaldehyde [$C_6H_{13}CHO$], and acetaldehyde [CH_3CHO] [*B7*]. He continued this research until 1873.

Borodin wrote to Butlerov in Kazan about his activity during this period of his life:

> I read organic chemistry in the academy (three lectures a week), and besides this I was entrusted with participating in the organization of the laboratory. They said that since nine-tenths was already done, designed, and firmly established, there would be nothing for me to do. But there is a world of trouble with engineers and interference by various people, all of which has forced me to become inactive in the work of the laboratory. Thanks, however, go to Zinin, who was constantly on my side, and together we were able to think of something and to arrange for maintaining work in the laboratory. Construction of this building, due to various transactions and tricks of the contractors with the engineers and the scandalous ignorance or unconscientiousness of the latter, was shamelessly dragged out. When I returned from abroad, I had hoped to find everything finished. However, only the walls were ready. Then they delayed us with promises that the laboratory would be ready in one month, in two months, in two and a half months, etc. We constantly met unforeseen delays, troubles, procrastinations; so for an entire year I remained without any possibility of working. Doing some trifles with aldehydes, I obtained some substances which, however, I have not been able to study intelligently, partly because of the lack of time and partly because of the impossibility of working efficiently in our old laboratory. I do not intend to describe the new laboratory or the old one to you, for Markovnikov[5] saw both of them and will report to you far better orally. The laboratory will be good, but what can we do if we do not increase the staff? Day after day Zinin pleads with all his might, but it seems to be to no avail. They even cut down to an impossibility the number of attendants, which was scanty to begin with. They left only three, and there are one hundred and four places in the laboratory. The number of workers will certainly reach fifty (this was the condition without which they would not agree to give us money for the laboratory). They will not agree to give a special assistant for this work, and without one it is impossible. Although Zinin and I both insist that one and the same professor cannot be occupied both with reading the lectures and with the practical work of the students, nevertheless, nothing is allowed . . . Close to 2000 [rubles] a year is allowed for maintenance of the laboratory, besides some raw materials (acids, soda, potash, alcohol, and others) which are obtained from the chemist's shop. This is very little for a large number

of occupants, for it is impossible to deduct anything for broken dishes and materials. The students have nothing, they do not pay for the laboratory and do not receive a salary . . . It is impossible to superimpose on the students the obligation of any fee, for even without this, they pay 50 rubles a year to the academy.

All this greatly complicates future work in the laboratory and causes a somewhat unpleasant feeling, which already appeared earlier in the conviction of the impossibility of conducting business in the manner one would like. But in the future perhaps the higher authorities will be convinced that our demands were completely well-grounded. I have received yet another position. I have been invited to the Forestry Academy where, it is true, there are many lectures, but on the other hand, they are read during only one semester (winter) . . . Farewell, dear Aleksandr Mikhaĭlovich, I am sad that I can report nothing chemical to you but what can one do[?][6]

In 1863 the Conference of the Academy granted Borodin an apartment in the building of the Natural Science History Institute, which opened on October 13th. Borodin arrived here with Protopopova, whom he had married shortly before this[7]. Dobroslavin reported,

The new laboratory, opening in 1863 on the corner of the Aleksandr Bridge, is under Borodin's supervision as director since Zinin, its recent director, obtained an apartment at the laboratory in the Academy of Sciences, at about this time. He himself [that is, Borodin. — A.], just married to Ekat. Serg. [Ekaterina Sergeevna] Protopopova, moved into the new building, the first entrance from the Neva, and lived there until his death[8]. The laboratory was located in one corridor from his apartment, and Borodin worked there tirelessly with the students every day. But during his work Borodin always retained his fresh and good-humored disposition with his pupils and his colleagues and was always ready to interrupt any of his own work without impatience, without irritation in order to answer any questions asked of him. Workers in the laboratory felt as if they were in a family circle. But he did not forget music. While working, he almost always hummed something to himself; he readily talked and argued with the co-workers about musical novelties, trends, and techniques of musical composition. When he was in his apartment, we often heard the harmonious sounds of the professor's piano, which were carried along the laboratory corridor. Borodin's good humor and kindness affected everyone. Anyone could go to him with his ideas, questions, and views; there was never any refusal, haughty reception, or disregard. Very rarely were outbursts of irritation aroused in Borodin and then only by carelessness or a negligent attitude toward work by the workers in the laboratory. "Ah, old chaps," was then heard, "what are you doing? You will ruin all the instruments in the cupboard. Is it necessary to fill the air of this clean laboratory with all kinds of rubbish? Go into the black" [a special room for large-scale chemical work].

Borodin's warm, sincere attitude toward the students was not restricted to the laboratory. Almost all the workers there were accepted into his family as the closest of acquaintances. They often ate breakfast, dinner, and even supper with him, when they remained in the laboratory for a long time. Borodin's apartment was, it must be said, constantly open wide for all the young men. He was constantly troubled about the fate of each student on graduating from the academy, making use of every effort in order to give them help. It was often said of him that it was impossible to meet him in society without his asking about someone or arranging something for someone[9].

Sociable, with an exceptionally sympathetic character, Borodin maintained his acquaintance with old comrades. Professors of the Medical-Surgical Academy, Zinin, Sechenov, Botkin, Yunge, Sorokin[10], Koshlakov[11], Gruber[12], and others, whose names occupy an honored place in the history of Russian science, often visited Borodin. He himself often visited them.

Borodin wrote to his wife:

The visiting day has begun for me. Smol'skiĭ came, with whom I talked a little about his work, then Kashevarova[13], then Sorokin, then Mikhail Vasil'evich Uspenskiĭ[14], also a dozen and a half of all sorts of students, pharmacists, students preparing for doctor's degrees, etc., unknown neither to me nor to you. I talked with them until four o'clock. Yes! I forgot! Nikolaĭ Nikolaevich Zinin and Nikolaĭ Nikolaevich Lodyzhenskiĭ[15] visited me. Zabelin[16] and Sorokin are extremely dissatisfied with their foreign trip and say that now they would not go there for the world. "It's a frightful bore," according to Sokolov's expression[17].

As already mentioned, even abroad the idea of founding a Russian chemical society occurred to the Russian chemists who were members of the Heidelberg Circle. "Here" [in Heidelberg. — A.] Borodin wrote to Alekseev, "in the meantime a chemical society, domestic for the present time, has been established (from the beginning, of course, only within our own group)"[18].

Upon their arrival from abroad Russian chemists, including Borodin, organized in St. Petersburg a chemical circle which met the first Thursday of each month. Here scientific reports were presented, and issues and the newest trends in chemistry were discussed. This group was the forerunner of the Russkoe khimicheskoe obshchestvo (Russian Chemical Society) at St. Petersburg University, which was organized in 1868.

Footnotes to Chapter 6

1 V. V. Stassov: Borodin, p. 149. — A.

2 Borodin: Letters, I, p. 56. — A.

3 Mavrikiĭ Osipovich Vol'f (Wolf) (1825–1883) of St. Petersburg was a Russian bookseller and publisher. — T.

4 Borodin: Letters, I, p. 57. — A.

5 Vladimir Vasil'evich Markovnikov (1837–1904) earned his doctorate in 1869 from Kazan University, where he succeeded Butlerov in the Chair of Chemistry. He is best known for his rules governing addition reactions with unsymmetrical olefins. — T.

6 Letter of A. P. Borodin to A. M. Butlerov, July 15, 1863, Archives of the Academy of Sciences, USSR, Stock 22, Inventory 2, No. 21, published here for the first time. — A.

7 They were married on April 17, 1863 in St. Petersburg. — T.

8 The building was located at the corner of Nizhegorodskaia Street and Pirogovskaia Embankment. The apartment was on the first floor. — T.

9 V. V. Stassov: Borodin, pp. 149–150. — A.

10 Ivan Maksimovich Sorokin (1833–1919) was a specialist in the field of forensic medicine. He was a professor at the Medical-Surgical Academy beginning in 1863. — A.

11 Dmitriĭ Ivanovich Koshlakov (1835–1891) was a therapeutist and Professor of Internal Diseases at the Medical-Surgical Academy. — A.

12 Ventseslav Leopol'dovich Gruber (1814–1890) was an anatomist, a professor at the Medical-Surgical Academy from 1847, and a friend of Borodin's. — A.

13 Varvara Aleksandrovna Kashevarova-Rudneva (1847–1899) was one of the first woman doctors in Russia. — A. [According to the GSE, she was "the first woman in Russia to receive the title of physician and the degree of doctor of medicine." See GSE, 1976, Vol. 11, p. 472. — T.]

14 Mikhail Vasil'evich Uspenskiĭ was a physician. — A.

15 Nikolaĭ Nikolaevich Lodyzhenskiĭ (1843–1916) was a [diplomat and] composer-dilettante. — A. [Described by Rimsky-Korsakov: (My Musical Life, 2nd revised English edition, Alfred A. Knopf, New York, 1928, p. 70) as "an erstwhile wealthy landowner gone to ruin, . . . a young man of education, queer, easily carried away and endowed with a strong, purely lyric talent for composition, and a fairly good piano technique in the performance of his own compositions," he joined the Balakirev music circle (See Chap. 7) during the 1866–67 season, but his official duties often kept him abroad, and his presence at their gatherings was a rarity. See Jennifer Spencer: Nikolay Nikolayevich Lodizhensky. In: Grove, Vol. 11, p. 122. — T.].

16 Ivan Egorovich Zabelin (1820–1908). — T.

17 Borodin: Letters, I, p. 105. — A.

18 Letter of A. P. Borodin to P. P. Alekseev, May 18, 1861, published here for the first time. — A.

Chapter 7

"The Mighty Little Group"

The great composer Mikhail Ivanovich Glinka was the founder of Russian classical music. Indeed, his opera *Ivan Susanin*[1] is an unsurpassed model of the Russian heroic opera, in which for the first time in the history of music the people and national heroes are the characters. No less remarkable are his symphonic compositions, famous songs, and also the opera *Ruslan and Liudmila*[2], in which the beautiful music of Glinka harmoniously combines with the forms of the brilliant Pushkin poetry and famous novels.

Turning to the life of the people, to the folk musical creations, Glinka, by his own musical creations, laid the foundations of a genuine national music, on the basis of which, as Stassov wrote, Russian music grew and developed "into wonderful original beauty, talent, and force"[3]. Despite the fact that Glinka's operas were not cordially received in the "high society" of Czarist Russia, they acquired enormous popularity in the nation and in the course of more than 100 years never left the stages of opera theaters.

During the 1860's Glinka's traditions, in the cause of the struggle against servility to an alien culture, were selected and developed by a group of Russian composers, whose ideological-esthetic views in many cases were close to the opinions of Herzen, Belinskiĭ, Chernyshevskiĭ, and Dobroliubov.

During one of the evenings at S. P. Botkin's home in the autumn of 1862, Borodin became acquainted with Miliĭ Alekseevich Balakirev. At that time Balakirev was one of the most brilliant followers and propagandists of Glinka's work. An untiring champion for the originality of Russian music, Balakirev gathered around himself the young, talented forces, who were gifted to support the people's democratic, truly national Russian art against Western trends. Among Balakirev's friends were Mussorgsky, Rimsky-Korsakov[4], and Cui[5].

The small circle of Russian composers which Borodin joined was called the Balakirev Circle. In an 1867 newspaper article devoted to a concert in honor of the arrival of Slavic delegations, Stassov wrote about the "small but already mighty group of Russian musicians." This accurate expression of Stassov's later came into common use.

Appearing as champions of national character in art, the composers of the "Mighty Little Group" (Moguchaia kuchka) reflected in their works the ideas of the prominent social movement of the 1860's, which fought against serfdom and oppression of the masses.

The struggle for a national character in music was the fundamental aspiration that unified the Mighty Little Group. The demand for nationality in form, the adoption of subjects chiefly from historical Russia and the contemporary national mode of life, and the support of the folk song as the main source of musical form are the most important foundations of the artistic program and concrete creative practice of the members of the Mighty Little Group.

"Glinka's successors," Stassov wrote, "followed him, guided by his extraordinary example and initiative"[6].

From his very first meeting with Borodin, Balakirev appreciated the complete originality of his rich talent. Balakirev wrote to Stassov,

> Our acquaintance had a special significance for him in that before meeting with me he considered himself only an amateur and did not attach any significance to his efforts at composition. It seems to me that I was the first person who told him that his real work was composition[7].

At this time Borodin seriously began to consider his activity as a composer. Personal contact with Balakirev and Balakirev's friends and attendance at concerts of the Free Music School (founded in 1862 by Balakirev and Lomakin[8]), at which the best creations of Russian composers (chiefly those of Glinka and Aleksandr Sergeevich Dargomyzhskiĭ) were performed, contributed in no small part to Borodin's musical development. Stassov wrote,

> They met, spent the whole evening together with music, and of course they had a strong effect on one another, increasing the development of the spirit and fanning the poetic flame[9].

Borodin's close friend Rimsky-Korsakov, later recalling his first meeting with Borodin, wrote:

> At the very first of my visits with Balakirev I heard that a new member had appeared in his group, for whom he had high hopes. This was A. P. Borodin. When I moved to [St.] Petersburg, he had not been there for some time; he did not return until after summer. Balakirev played fragments of the first part of his E Flat Major Symphony[10] for me, which amazed rather than pleased me. Borodin returned shortly afterward. I became acquainted with him, and our friendship began at this time, although he was ten years older than I . . . At that time Borodin was already Professor of Chemistry in the Medical Academy and lived near the Liteĭn'yĭ Bridge in the academy's building, subsequently remaining in the same apartment until his death. My symphony[11], which Balakirev and Mussorgsky played for him in a four-hand arrangement, pleased Borodin. The first part of his symphony in E Flat Major was not finished, but he already had material for the other parts, composed by him during the summer abroad. I was delighted by these fragments, I also understood the first part, which simply amazed me at first acquaintance[12].

During the autumn of 1862 Borodin met Mussorgsky for the third time at Balakirev's home. Borodin wrote:

> Mussorgsky and I again recognized each other at once; we recalled both of our previous meetings. Mussorgsky already had greatly grown musically. Balakirev

wanted me to become acquainted with the music of his group and first of all with the symphony "of the absent one" (It was Rimsky-Korsakov, at that time a naval officer, who had just left for a long voyage to North America). Mussorgsky sat with Balakirev at the piano (Mussorgsky the primo, Balakirev the secondo). The playing was certainly not that of the first two meetings. I was struck by the brilliance, the sensibility, and the energy of the performance, and the beauty of the work. They played the finale of the symphony. Mussorgsky knew that I had some feeble impulse to write music so he began to ask me to do something. I was terribly embarrassed and refused point-blank[13].

Under Balakirev's first-hand guidance Borodin began to write his first symphony. However, the composition progressed slowly. It was extremely difficult for Borodin to find free time for music. From 8:00 in the morning until evening he worked in the chemical laboratory. A large number of different complex responsibilities in various institutions occupied almost all of his time.

During the evenings colleagues from the academy or musical friends visited Borodin, and only at night, when in complete silence by himself in the office, did he surrender himself to musical work. There were days when Borodin gave up music completely. With a touch of his habitual humor he wrote:

Music sleeps; the altar of Apollo goes out; the ashes on it get cold; the muses weep, near them urns are filled with tears, tears flow over the brim, they flow together into a stream, the stream babbles and with sadness tells of my cooling to the arts today[14].

Rimsky-Korsakov recalls,

Borodin was to the highest degree a sincere and well-educated person, a pleasant and distinctively writty conversationalist. Visiting him, I would often find him working in the laboratory which was located next door to his apartment. One time when he was sitting over the flasks which were filled with some kind of colorless gas and was distilling it by means of tubes from one container to another, I told him that he was "transferring emptiness into vacancy"[15]. When he finished his work, he would enter his apartment with me, and we would play together or talk. Right in the middle he would leap up and rush back into the laboratory in order to observe if something had burned out or boiled over, all the while making the corridor echo with incredible sequences of successive ninths or sevenths. Then he would return, and we would continue the music or the interrupted conversation[16].

Borodin's first symphony was finished in 1867. It was performed for the first time on January 4, 1869 in St. Petersburg under Balakirev's direction with great success.[17] This symphony already displayed the characteristic traits of Borodin's works – epic power, sincerity of lyrics, and a strong bond with the national-folk foundations of Russian music, which are invariably observed both in the expansive melodiousness of the melody and in the originality of the rhythms.

The performance of V. A. Krylov's[18] play-parody *Bogatyri* (The Valiant Knights) took place on November 6, 1867 in the Bolshoi Theater in Moscow for the benefit of the stage manager of the theater, N. P. Savitskiĭ. The music was composed by Borodin and satirically ridiculed the convention of the genre of "the grand opera"[19].

The works of the Balakirev Circle flourished during the 1860's. During the period from 1867 to 1872 Mussorgsky wrote *Boris Godunov*, Rimsky-Korsakov wrote *Pskovitianka* (The Maid of Pskov), and Cui wrote *Ratcliff*.[20] Borodin commented to his wife,

> Cui acquainted me with all the new numbers of "Ratcliff" and with the orchestration of the whole opera. Korsin'ka [Rimsky-Korsakov] played several numbers from his opera "Pskovitianka." Now I tell you that such fragrance, such youth, freshness, and beauty make me simply become limp with pleasure. What a tremendous talent this man has! And what ease of creation! Then Mussorgsky performed the first act of "Zhenit'ba" [The Marriage], which is based on Gogol's text[21].

During these same years Borodin's talent developed to complete brilliance. At the end of the 1860's he wrote remarkable songs to his own words: the song-tale Spiashchaia kniazhna (The Sleeping Princess), dedicated to Rimsky-Korsakov (1867), Pesnia temnogo lesa (Staraia pesnia) (The Song of the Dark Forest) (Old Song), Fal'shivaia nota (The False Note), Morskaia tsarevna (The Sea Princess), Otravoĭ polny moi pesni (My Songs are Filled with Poison)[22] to the words of Heine, and the ballad More (The Sea), dedicated to Stassov (1870).
Stassov wrote,

> The song "The Sea" is the best of all Borodin's songs and, without doubt, one of the most talented of those now existing in the world. However, it is not from the text which he had originally intended. The present text speaks only of "a young swimmer, carrying with him expensive goods. With a rich booty he goes home; with colored gems, with expensive brocades, with large pearls, with a purse of gold, and with a young wife." Originally another image was planned. That very music, which we now know, pictured a young exile, involuntarily leaving his native land for political reasons. Full of impassioned and burning hopes, he returns home only to meet a tragic death in a storm within sight of the shores of his native land[23].

Borodin himself wrote of this song:

> The piece turned out well: much animation, fire, and sparkle, and it is tuneful as well. Everything in it "rings true" from a musical point of view[24].

Stassov wrote,

> Some of these songs, as, for example, "Spiashchaia kniazhna" (Sleeping Princess), "Morskaia tsarevna" (The Sea Princess), "Pesnia temnogo lesa" (The Song of the Dark Forest),[25] were full of deep and mighty epic spirit, as if from one of the better pages from Glinka's "Ruslan." That very form and outline, which were necessary to be silhouetted one day in the opera "Prince Igor" by wonderful poetry and strength, appear here[26].

During these years Borodin's contemporaries clearly sensed the national originality of his work and its continuity with the music of Glinka. In 1867–1868 Borodin wrote some choruses for the opera Tsarskaia nevesta (The Czar's Bride) on the subject of L. A. Meĭ's drama, suggested to him by Balakirev. However, this project was not finished. Rimsky-Korsakov later wrote an opera on this subject.

One of the creators of original Russian folk music, which constantly resounded in concerts arranged by the Free [Music] School and in impromptu concerts in the group of his friends, Borodin gravitated most of all toward the epic repository of music. This is evident in his treatment of Slovo o polku Igoreve (The Lay of Igor's Campaign), which served as the subject for *Prince Igor*, the first Russian heroic-epic opera to appear as the immediate sequel to the tradition of Glinka.

The Mighty Little Group placed itself at the head of the heroic struggle against reactionary, antinational tendencies in Russian music. Creating great new music, the Balakirev Circle, following Glinka and Dargomyzhskiĭ, encouraged the blossoming of Russian classical music, whose ideological sources were the people's real national creation. In the relentless struggle with conservative, reactionary tendencies, which defended the thesis "art for art's sake," the Mighty Little Group became the guiding star for everything progressive and revolutionary in Russian music. Faith in the strength of the Russian people and the aspiration to show truthfully the growth of its national consciousness are the traits which characterized the works of the members of the Mighty Little Group.

The reactionary leaders of the Imperatorskoe Russkoe Muzykal'noe Obshchestvo (Imperial Russian Musical Society) (founded in 1860) [by Anton Grigor'evich Rubinstein — T.] attempted to create opposition to the new independent Russian musical school in order to slow down and to destroy the movement which had been started.

At a time when those reactionaries who fawned before the West were raving about the new musical trend, the classical productions of the members of the Mighty Little Group won the ever-growing popularity and love of the Russian people. Borodin wrote to his wife,

> Balakirev and all of the group were defamed and not spared the most foul swearing and the most viele slander. Three of the season's numbers were exclusively dedicated to the torrent against Miliĭ [Balakirev. — A.]. But, as if to spite them, Miliĭ's reception in each concert was warmer and warmer. And today the public simply received Miliĭ enthusiastically and called for him many times after "Elizaveta", after "Sadko", and after the symphony. This reception serves as the best answer to the insults and slanders of obscurants and the Mikhaĭlovskiĭ Palace with its vile minions[27].

During the winter of 1868–1869 Borodin wrote several critical articles in the Sankt-Petersburgskie Vedomosti (The St. Petersburg News) about compositions which were performed in concerts of the Imperial Russian Musical Society and the Free Music School. Deeply loving his people and the rich Russian folk music, Borodin, with his characteristic warmth, was exasperated by the narrow-minded "fraternity" of the musical society who worshipped the music of the West. In a letter to Balakirev, Borodin wrote,

As far as I am concerned, I am disappointed in our fraternity ... Each one strives to shrink to a Frenchman or an Englishman, to fawn before the judgment of Europe. There is not the smallest display of national independence; rather, there is a complete lack of individuality.[28]

There is an interesting letter from Borodin to his wife in which he permits himself to express his thoughts about the untalented guardian of the Musical Society, the Grand Duchess Elena Pavlovna[29]:

Using a nonpostal means of communication, I will tell you something about a person of a most august house who has such a lively regard for musical affairs in Russia. When Cui wrote one of his typical satires[30], where the significance and the recent state of the Musical Society were touched upon, Elena Pavlovna flew into an indescribable rage. Having marked in pencil all that she found revolting in this satire, she sent messengers directly to her toadies. Because of this circumstance a messenger from the manager of the public library, Baron Korf',[31] suddenly appeared that evening at the home of Stassov, his subordinate. Korf' demanded to see Stassov immediately on important business, but if it were impossible to come at once, then tomorrow morning. In the morning a new messenger appeared for a double confirmation of his superior's request. Stassov went to Korf'. With a haughty and outraged look Korf' extended two fingers to him and reported that the Grand Duchess commanded him to convey her most extreme displeasure with Stassov's article in which he dared to insult Her Highness in the most impudent manner. Stassov, who was in no way guilty of Cui's article, stared at Korf' and said: "For goodness' sake, am I to blame for this? Did I really write this?" It turned out that the old dame in her rage had made a mess of things, not even finding out who the guilty party was. Korf' was unspeakably glad and uttered a scream: "Well, thank God, we are saved!" Indeed, Europe is saved! So they announced to the Grand Duchess that Stassov did not write it, but Cui. Then a General from the Grand Duchess went to Totleben[32] with the same satire. Totleben was ordered to read the terrible satire and reprimand his subordinate, that is, Cui, most severely. Totleben was never in his life particularly interested in music, but he had to read Cui's article, and try as he could, he could not find anything insulting in it. But there was nothing else to do so he called Cui's closest superior. The superior called Cui himself and commanded him to set off for Totleben. The latter reported all reprimands according to the Grand Duchess' order, but then in addition he begged Cui to continue writing satires (carefully, of course) so that no one will think that the chief forbids Cui to write. This is a tempest in a teapot. And she, the old dame, should be ashamed. You see, she's making a fool of herself[33].

Be that as it may, the action of "Elena and company" affected the activity of the Free Music School. Borodin wrote to his wife: "For the sake of the concerts, ... [Balakirev] had to run, solicit, search, and support the various alliances which were necessary for the struggle with Elena and company"[34]. "The person of the most august house," making use of her position, did not stop at anything in order

Fig. 8. Aleksandr Porfir'evich Borodin (1873)

to discredit and humiliate the growing name of the genuine folk productions of the leading figures of Russian music. Realistic creative productions of the Mighty Little Group members could not but call forth a hostile attitude in circles of reactionaries of all colors.

It is impossible, however, to consider the entire Russian Musical Society as a focus of reactionary ideas in music. Progressive Russian musicians also were affiliated with this society, making use of its concert stage for their performances. Speaking of the struggle of The Five with reactionary spokesmen of the Russian Musical Society, it is impossible not to mention those ties which existed between the works of the Mighty Little Group and the musical activity of the brothers, A. G. and N. G. Rubinstein[35] and especially of Tchaikovsky.

Tchaikovsky's works exerted a strong influence on the followers of the Mighty Little Group, who were united around Rimsky-Korsakov in the Balakirev Circle. It should also be noted that the Moscow section of the Russian Musical Society, founded by N. G. Rubinstein in 1860, was not associated with the reactionary circle of the Mikhaĭlovskiĭ Palace in St. Petersburg and was favorably disposed toward the works of the Mighty Little Group.

Footnotes to Chapter 7

[1] This opera brought Glinka his first fame. Originally titled *A Life for the Czar* and first produced in 1836 at St. Petersburg, it was renamed *Ivan Susanin* after the October Revolution of 1917. — T.

[2] First produced in 1842, Glinka's second opera, with its Russian, oriental, and "fantastic" conventions (whole-tone scale, etc.), lyrical melody, and colorful, transparent harmony and orchestration, provided models on which the composers of the Mighty Little Group formed their styles. — T.

[3] V. V. Stassov: Sobranie sochineniĭ (Collected Works) [hereafter abbreviated as Collected Works], [Tipografiia M. M. Stasiulevicha], St. Petersburg, 1894, Vol. 3, p. 763. — A.

[4] Nikolaĭ Andreevich Rimsky-Korsakov (1844–1908) was a most prominent Russian composer, a member of the Balakirev Circle, and a close friend of Borodin's. — A. [See Gerald Abraham: Nikolay Andreyevich Rimsky-Korsakov. In: Grove, Vol. 16, pp. 27–41. — T.]

[5] Cesar (Tsezar') Antonovich Cui (Kiui) (1835–1918) was a famous Russian composer, musical critic, and a member of the Balakirev Circle. He became acquainted with Borodin in about 1864. — A. [See Geoffrey Norris: Cesar Cui. In: Grove, Vol. 5, pp. 92–95 and V. V. Stassov: Tsezar' Antonovich Kiui, Biograficheskiĭ Ocherk, Gosudarstvennoe muzykal'noe izdatel'stvo, Moscow, 1954. — T.]

[6] V. V. Stassov: Collected Works, p. 651. — A.

[7] V. V. Stassov: Borodin, p. 153. — A.

[8] Gavriil Yakimovich Lomakin (1812–1885) was a teacher of choral singing. He was the director of the Free Music School until January 1868. Concerts of the Free School during the time of his directorship often were colloquially called Lomakin's Concerts. — A. [See Jennifer Spencer: Gravriil Yakimovich Lomakin. In: Grove, Vol. 11, p. 139. — T.]

[9] V. V. Stassov: Collected Works, Vol. 3, p. 765. — A.

[10] Borodin's Symphony No. 1, in E Flat Major, composed during the years 1862–1867 and first published in full score by Bessel in 1882, was dedicated to Balakirev. — T.

[11] Probably Rimsky-Korsakov's Symphony No. 1, in E Flat Minor, first performed with gratifying success in St. Petersburg on December 31, 1865, when the composer was only 21 years old. — T.

[12] N. A. Rimsky-Korsakov: Letopis' moeĭ muzykal'noĭ zhizni (Chronicle of My Musical Life) [hereafter abbreviated as My Musical Life, [Gosudarstvennoe muzykal'noe izdatel'stvo], Moscow, 1935, 5th edition, pp. 60–61. — A. [For an English translation by Judah A. Joffe see: My Musical Life, 2nd, revised edition, Alfred A. Knopf, New York, 1928, pp. 52–53. — T.]

[13] V. V. Stassov: Borodin, p. 152. — A.

[14] Borodin: Letters, I, p. 62. — A.

[15] The Russian expression for "chewing the rag" or the useless work of the Danaides of Greck classical mythology. — T.

[16] Rimsky-Korsakov: My Musical Life, p. 61. — A. [English version, p. 53. — T.]

[17] In his 1963 biography of Borodin, S. A. Dianin states that the antagonists of the Balakirev Circle refused to acknowledge the success of this symphony "of a composer seeking oblivion," as Borodin called it, as witnessed by a remark of Serov's: "A symphony by someone called Borodin gave little satisfaction. Only his friends called and applauded him with any enthusiasm." [From the newspaper Golos (The Voice), 1869]. — T.

[18] Viktor Aleksandrovich Krylov (1838–1906) was a writer, dramatist, and theatrical agent. [See GSE, 1976, Vol. 13, p. 531. — T.] For information on Bogatyri (The Valiant Knights) see the work of Igor Glebov (Academician B[oris] V[ladimirovich] Asaf'ev): Iz zabytykh stranits russkoĭ muzyki (From the Forgotten Pages of Russian Music), Muzykal'naia letopis' (Musical Chronicle), Collection I, 1922. — A.

[19] The performance was not a success; both the critics and the public failed to understand that the music was intended as a burlesque. — T.

[20] Based on Heinrich Heine's tragedy William Ratcliff. — T.

[21] Borodin: Letters, I, pp. 108–109. — A.

[22] During the summer of 1868 Anna Nikolaevna Kalinina (née Lodyzhenskaia — she was the sister of the composer Nikolaĭ Nikolaevich Lodyzhenskiĭ — see Note 15, Chap. 6) fell in love with Borodin, who told his wife of the affair in a letter of October 25, 1868: "My feelings toward her do not alter the way I feel toward you, and I am giving only that which I cannot give to you; it is nothing more than that 'feeling of mine toward children,' in her words, toward weakness, youth, hopes, and the future . . .". Anna's influence led Borodin to compose "The Sea Princess," to his own words (dedicated to A. E. Makovskaia, published by Bessel in 1873), in July 1868, and later "My Songs Are Filled with Poison" (words by Heinrich Heine; dedicated to Cesar Cui; published by Jurgenson in 1870) and "The False Note" (words by Borodin; dedicated to Modest Petrovich Mussorgsky; published by Jurgenson in 1870). These three songs reflect his state of mind at the time and reveal the tension in his relations with Ekaterina Sergeevna. — T.

[23] V. V. Stassov: Borodin, p. 156. — A.

[24] Borodin: Letters, I, p. 200. — A.

[25] According to A. P. Dianin, the censor found The Song of the Dark Forest "seditious" and was not going to pass it. Rimsky-Korsakov therefore placed it between two of his own songs, which were innocuous. The censor, accustomed to passing Rimsky-Korsakov's songs without even looking at them, passed it on June 13, 1873. — T.

[26] V. V. Stassov: Borodin, p. 156. — A.

[27] Borodin: Letters, I, p. 168. — A.

[28] Ibid., p. 94. — A.

[29] Grand Duchess Elena Pavlovna (1806–1873) [née Frederika Charlotte Maria of Württemberg] was the [founder and] "avgusteĭsheĭ pokrovitel'nitseĭ" (Most August Protectress) of the Russian Musical Society and imagined herself to have musical taste. Actually, she often supported various untalented rascals and musical reactionaries. — A. [The wife of Grand Duke Mikhail Pavlovich, she was the hostess of one of the most influential salons in St. Petersburg in the 1850's and early 1860's. — T.] Mussorgsky ridiculed her in his Raĭk (Peep Show) under the name Muse Euterpe. — A. [See W. Bruce Lincoln: Elena Pavlovna. In: Joseph L. Wieczynski (editor): The Modern Encyclopedia of Russian and Soviet History, Academic International Press, Gulf Breeze, Florida, 1979, Vol. 10, pp. 171–175. — T.]

[30] He had in mind the satire Muzykal'nye zametki (Musical Notes) signed "+ + +" and found in No. 304 of the Sankt-Peterburgskie vedomosti (St. Petersburg News) of November 4, 1870. The witty narrative about the organization "of the Committee of the People's Defenses" in the Russian Musical Society filled Elena Pavlovna especially with indignation. — A.

[31] Modest' Andreevich Korf'. See V. V. Stassov: Graf' Modest' Andreevich Korf' 1800–1876, Tipografiia V. S. Balasheva, St. Petersburg, 1876. — T.

[32] Eduard Ivanovich Totleben (1818–1884) was a military engineer and a hero of the defense of Sevastopol. — A. [See GSE, 1981, Vol. 26, p. 254. — T.]. He was the author of Opisanie oborony g. Sevastopolia (Description of the Defense of Sevastopol). — A. [Tipografiia N. Tiblena, St. Petersburg, 1863, which was translated into German and French. — T.]

[33] Borodin: Letters, I, pp. 271–272. — A.

[34] Ibid., p. 235. — A.

[35] Anton Grigor'evich Rubinstein (1829–1894) was a noted pianist and composer. He conducted the concerts of the Russian Musical Society until 1867. — A. [See Edward Garden: Anton (Grigor'yevich) Rubinstein. In: Grove, Vol. 16, pp. 297–300. — T.] Nikolaĭ Grigor'evich Rubinstein (1835–1881) was a well-known pianist, conductor, and composer. He was founder and first director of the Moscow Conservatory. — A. [See Edward Garden: Nikolay (Grigor'yevich) Rubinstein. In: Grove, Vol. 16, pp. 301–302. — T.]

Chapter 8

The Peak of His Scientific Work

Despite his enthusiasm for music, Borodin continued to work intensively in the chemical laboratory. The completion of the construction of the new chemical laboratory of the Medical-Surgical Academy, which was abundantly equipped for that time, made it possible for him to start complex research on the condensation of aldehydes, which led him to the discovery of aldehyde resins, aldol, and many other extremely interesting chemical compounds.

On April 15, 1864 Borodin was elected Ordinary Professor [Full Professor. — T.] in the Department of Chemistry of the Medical-Surgical Academy[1].

In May of the same year he completed the work Issledovaniia o deĭstvii natriia na valeral'degid (Investigations of the Action of Sodium on Valeraldehyde) [*B 14*]. In this work Borodin studied the products of the interaction of aldehydes with alkali metals for the first time. At this time chemists thought that aldehydes, like alcohols, were able to form metal derivatives by substitution of hydrogen by a metal. By his research Borodin showed that:

1. By the action of sodium on valeraldehyde a substance is formed which is not a simple product of the substitution of hydrogen by the metal but which consists of a complex mixture of various chemical compounds.
2. Neither valeraldehyde nor substances that are isomers or polymers of it are among the reaction products.
3. The product $C_{10}H_{22}O$, obtained from this reaction, appears to be a monohydric alcohol, which may be identical with or an isomer of the alcohol corresponding to "isocapric" acid.

The significance of this research on the condensation of aldehydes lay in the fact that the new experimental results allowed Borodin to work out a general method for the condensation of aldehydes in the presence of metallic sodium and potassium and also of caustic alkalis.

In addition to his scientific work in the chemical laboratory during the winter Borodin read the lectures on organic chemistry (three times per week) to students of the academy, and during the spring he supervised examinations, attended numerous conferences, and participated in various committees.

On December 28, 1867 the First Congress of Russian Naturalists and Physicians was inaugurated in St. Petersburg. Borodin participated in the work of the Congress as a member of the chemical section. At this Congress the question of establishing a Russian Chemical Society was posed by the chemists.

Borodin wrote to Alekseev,

As you well know, we have finally inaugurated a Chemical society, and the first meeting was at the beginning of December [1868. — A.] in the university auditorium. Zinin was elected president, Menshutkin[2], secretary, and [G. A.] Shmidt, treasurer. It was very gay and pleasant. The second meeting is set for January 9, 1869. What can I say about my works?

I have still obtained nothing sensible with the polymers of valeraldehyde. You cannot touch it; either there is no action at all or else removal of the elements of water n $C_5H_{10}O$—m H_2O occurs, and products are obtained which are very high boiling and highly stable and with which nothing can be done. I wanted to drive off the removable water to [obtain] the hydrocarbon C_5H_8 ($C_5H_{10}O$—H_2O). This would be a homolog of acetylene,[3] an interesting substance, but this reaction has still not been achieved; I would be happy to obtain at least the polymer $C_{10}H_{16}$, a commonplace hydrocarbon, but I do not even obtain that. By the way, I undertook work of another sort, that of the attraction of water by deliquescent and hygroscopic substances and the evaporation of their aqueous solutions under various conditions of temperature and pressure. This work has turned out to be frightfully difficult beyond all expectations, especially concerning the condition of temperature. I do not yet know what will result. In addition, of necessity, I was then busy with Liebreich's[4] protagon[5]. Such nonsense is found in the decomposition products that even the devil himself cannot make it out. One is neurine – a sensible substance[6].

On January 9, 1869 at a meeting of the Chemical Society Borodin gave a report on the products which he obtained by the action of bromine vapor on the silver salts of valeric and butyric acids [*B 18*].

As early as 1860, as we have seen, Borodin attempted to prepare monobromovaleric and monobromobutyric acids by the action of bromine vapor on the silver salts of the corresponding organic acids [*B 7*]. By doing this he actually succeeded in obtaining products of the substitution of silver by bromine, one of which he analyzed. The analysis showed that the composition of the product corresponded to the composition of monobromovaleric acid. Borodin's attempt to prepare salts of this acid in the pure form met with little success since these products are not very stable.[7]

Slightly more than 30 years later the well-known Russian organic chemist Gustavson[8] wrote to his friend, the chemist Dem'ianov[9], that the production of the bromo-substituted derivatives of organic acids, developed by Borodin,

now proves to lead to a very original general method for obtaining esters, which, of course, should be expanded and developed as far as possible[10].

At the second Congress of Naturalists [and Physicians], which took place in Moscow on August 20–30, 1869, Borodin presented a new report on his discovery of a new fatty acid, "isocapric" acid, $C_{10}H_{20}O_2$;[11] salts of this acid; and its aldehyde, $C_{10}H_{20}O$.[12] These products were obtained from "isocapric" alcohol, $C_{10}H_{22}O$,[13] which was synthesized by Borodin by the action of sodium on valeraldehyde.[14] Borodin described in detail the properties of and the method for obtaining these "isocapric" compounds [*B 19*].

On October 2, 1869 at the Russian Chemical Society, Borodin gave a preliminary report on his research in progress on compounds which were obtained by the removal of water from aldehydes [*B 20*]. Borodin obtained these compounds in part by heating the aldehydes in a sealed tube and in part by the action of a solution of potassium hydroxide on sodium hydrogensulfite [$NaHSO_3$] compounds of the aldehydes. From valeraldehyde he obtained two compounds, one of which was distilled at a temperature of from 195° to 200°, the other at about 300°. Two compounds were also obtained from enanthaldehyde.

In letters to his wife [September 21, 1869. — T.] during this period Borodin described in detail his chemical and musical activity:

> I worked all day in the laboratory. I am now in a completely happy period of laboratory activity. Everything is going well, and I am now in the passion of laboratory work. An unpleasant clash with Kekulé almost occurred on the chemical field. In one of his works he touched that field in which I am working. It is true that he proceeded from completely different principles and did not touch on my topic, but nevertheless, in a subsequent move of his research he could easily come across and pursue my very ideas. To prevent the possibility of a clash, yesterday I reported my work at a meeting of the Chemical Society although it was still far from completed. However, all the

Fig. 9. Members of the Chemical Section, 1st Meeting of Russian Naturalists (St. Petersburg, January, 1868). Seated (left to right): V. Yu. Rikhter, S. I. Kovalevskiĭ, N. P. Nechaev, V. V. Markovnikov, A. A. Voskresenskiĭ, P. A. Il'enkov, P. P. Alekseev, A. N. Engel'gardt. Standing (left to right): F. R. Vreden, P. A. Lachinov, G. A. Shmidt, A. R. Shuliachenko, A. P. Borodin, N. A. Menshutkin, N. N. Sokovnin, F. F. Beilstein, K. I. Lisenko, D. I. Mendeleev, F. N. Savchenkov

chemists found it extremely interesting both from the practical side and from the theoretical solution of ideas. I have been busy these days from morning to evening with the literature collected on this subject. I have also supplied the musicians with the first number of "Igor"[15] where Yaroslavna's dream appears charmingly[16].

In another letter Borodin wrote [on March 9, 1870. — T.]:

Kekulé (in Bonn) reproaches me as if I had appropriated from him the work (i.e., not the same work from a practical side but the idea of the work) with valeraldehyde (which I am now doing). He published this in the Berichte of the Berlin Chemical Society[17]. Such a trick compelled me then and there to make a statement about the facts discovered by me and to show that I was occupied with these questions as early as 1865 but that Kekulé stumbled upon them only in August of last year. Look at this honest-minded German! Although our Chemical Society knew all about this, I considered it necessary to announce it so that it would be communicated by the usual procedure to the Berlin Society[18].

It is well known to historians of chemistry that this was not the first time that Kekulé tried to usurp the discoveries of Russian scholars. He tried to attribute to himself the priority for the creation of the theory of the structure of organic compounds, unquestionably belonging to the great Russian scholar Aleksandr Mikhaĭlovich Butlerov[19].

As is known, as early as 1861 Butlerov came forward at the [Chemical Section of the] Congress of German Naturalists and Physicians in the town of Speyer with the report in which he established the principles created by him of the theory of the structure of organic substances. This report was received with extraordinary coldness by German scientists although it contained a most important theoretical thesis, which lay at the foundation of the subsequent development of chemistry[20].

Several years after this report, when Butlerov's ideas were shown to be predominant in chemistry, Kekulé was not ashamed to announce his priority in the theory of the chemical structure of organic substances despite the fact that this clearly violated historical truth[21].

It fell to Butlerov to endure the struggle for the honor of Russian science. He was forced to go abroad in 1867–68 specifically to defend his discovery.[22] To restore the historical truth he published in German[23] [the second edition of — T.] his famous textbook: Vvedenie k polnomu izucheniiu organicheskoĭ khimii (An Introduction to the Complete Study of Organic Chemistry), already published in Russia [Kazan. — T.] in 1864.

In a request addressed to the Physical-Mathematical Faculty of Kazan University concerning the assignment of his mission abroad, Butlerov wrote:

During the time of my last five-month stay abroad in 1861, the article O khimicheskom stroenii veshchestv (On the Chemical Structure of Substances) was published and read by me at the Congress of German Physicians and

Naturalists. Since then I have constantly pursued the principle of chemical structure in all of my writings and works. Some of these investigations were stated in detail by me in the article Ob ob'iasnenii sluchaev izomerii (On the Explanation of Cases of Isomerism). And later it was taken as a principle in my textbook of organic chemistry. Facts discovered by me in the last years, and various discoveries of other chemists confirmed the soundness of my views . . . All this, however, does not prevent many foreign scholars from inadequately recognizing that part which I took in working out the views which are now predominant in science. And they often state, as new, what was already said by me . . . For the person who is dedicated to science, complete acknowledgment of other specialists of what actually has been accomplished by them constitutes one of the best rewards and the best encouragement for further activity. I must say that I highly appreciate such acknowledgment, but I consider a personal communication of my views to my fellow scientists as the best method for its achievement. I dare to anticipate that both the faculty and the Council will not regard the position of their fellow workers in the scholarly world with indifference[24].

At this same time Kekulé ungraciously tried to claim another important discovery of a Russian scholar, Borodin's discovery. Kekulé's assertion that the idea for the research on the condensation of aldehydes belongs to him forced Borodin to come forward, as we saw, with a corresponding refutation and later led to the controversy with Kekulé with respect to priority.

Borodin wrote to his wife [on March 20 or 21, 1870. — T.],

I made up my mind not to answer Kekulé but simply to continue my work. Otherwise he can think that I was really startled by his statement. When my work is finished, I will make a casual notice of Kekulé in passing. This is far more tactful[25,26].

In 1869 Butlerov moved from Kazan to St. Petersburg. He had already earlier been acquainted with Borodin from Heidelberg. Meeting at the sessions of the Chemical Society and at conferences, they became close and visited one another. Borodin wrote to his wife,

On Friday I was at Butlerov's. He earnestly invited me when we were at the Chemical Society on Thursday. He asked all about you and said that Nadezhda Mikhaĭlovna [Butlerov's wife. — A.] wanted to see you without fail and asked when it would be possible to visit you. On the next day, Friday, I set off for Butlerov's. He was terribly pleased to see me and she also. In my opinion she had not changed especially, only a little gray. Her Misha [Butlerov's son. — A.], whom I saw in Heidelberg as a small boy, has now already finished the course in the gymnasium[27]. We indulged in various recollections about life abroad, and no small part fell to you. Nadezhda Mikhaĭlovna made inquiries about all sorts of things concerning you. Knowing that Butlerov suffers from various heart attacks, I persuaded him to go to Botkin and proposed to go tomorrow, that is, on Saturday. On Saturday evening I was with Butlerov

Fig. 10. Aleksandr Porfir'evich Borodin, portrait by I[l'ia] E[fimovich] Repin

at Botkin's and sat there until 3 o'clock in the morning. It was very pleasant and gay at Botkin's[28].

The beginning of the 1870's was the time of the most interesting work in the chemical laboratory for Borodin. In letters to his wife he often reports:

> I lead a chemical life for the most part ... I am not occupied at all with music at this time. There is no time, there is absolutely no time ... I do not see my musical brotherhood for the time being, there is no time. I still have not seen Korsin'ka [Rimsky-Korsakov. — A.] ... In spite of a great number of pursuits, and indeed highly feverish ones, I had time to do the first number of Act I of "Igor" although it is not yet complete. I have not yet shown it to anyone[29].

On October 8, 1870, at a meeting of the Chemical Society, Borodin reported on the work, Sochetanie benzoïnogo al'degida s paranitroanilinom i nitrobenzoïnogo al'degida s anilinom i paranitroanilinom (The Combination of Benzaldehyde with Paranitroaniline and Nitrobenzaldehyde with Aniline and Paranitroaniline)[30], which was carried out by the student Lazarenko[31] under his guidance. On December 3rd of the same year Borodin reported Lazarenko's second work, Sul'fotsetenovaia kislota (Sulfocetenic Acid). Lazarenko, in the laboratory of the academy under the firsthand guidance of Borodin, obtained this acid by the action of sulfuric anhydride [sulfur trioxide, SO_3. — T.] on cetene[32]. The potassium salt of this acid was also obtained by him.

Organization of the laboratory work with the students took much of Borodin's time. Borodin wrote to his wife,

> I am now up to my neck with the organization of the laboratory. I surrender myself to the work of putting the laboratory equipment in order. I have lectures on Mondays, Tuesdays, Wednesdays, and Thursdays from 12 until 1:30.[33]

In another letter he wrote:

> I am overloaded with academic affairs because of the approaching New Year, by which time every account of the past year must be made, various business, committees, orders must be concluded, affairs of the laboratory must be eliminated, books must be presented, etc. . . . On account of all of this I almost never visit anywhere. By the way, now working with me is the wife of a student, only beginning, Lukanina,[34] who is already performing quantitative analyses and working judiciously.[35]

On February 1, 1871, at a meeting of the Chemical Society, Borodin gave a report on Doctor Krylov's work, Opredelenie zhirov (The Determination of Fats), which was carried out under his guidance.[36] They found that in determining fat in the muscles of the heart (upon fatty degeneration of the latter) a mass was obtained, resembling lecithin and containing phosphorus (5.39% P_2O_5) and nitrogen (1.57% N) mixed with the fat, which gave upon saponification not glycerin but cholesterol. This research threw new light on the chemical aspect of the phenomena of fatty degeneration of muscles.

On April 8th of the same year, at a meeting of the Chemical Society, Borodin announced that Lukanina, under his guidance in the chemical laboratory of the academy, had studied the oxidation of protein by potassium permanganate (chameleon). In this reaction, as would appear according to Béchamp's[37] data, urea is never a product; a substance is formed, which is rich in nitrogen, nonvolatile, soluble in water and alcohol, and which produces ammonia by the action of alkali.

A year later, at a meeting of the Chemical Society (February 3, 1872), Borodin presented a report on Lukanina's behalf, O deĭstvii khloristogo suktsinila na benzoin (On the Action of Succinyl Chloride on Benzoin). Lukanina, on Borodin's suggestion, investigated the action of succinyl chloride on benzoin and showed that succinyl dibenzoin, $(C_{14}H_{11}O_2)_2 \cdot C_4H_4O_2$, is obtained.[38] This substance crystallizes in rhombic plates at its melting point of 129°. In this manner Lukanina's investigations disproved Limpricht's[39] statement that succinyl chloride does not react with benzoin.

On May 4, 1872, at a meeting of the Chemical Society, Borodin presented three reports in which he gave an account of the results of his work of many years on the condensation of aldehydes [*B 23–25*]. Borodin studied the condensation of acetaldehyde, valeraldehyde, and enanthaldehyde. He did not study the condensation of formaldehyde, apparently because under ordinary conditions it is found in the gaseous state.

In the first report [*B 23*] Borodin gave an account of the results of the research on the action of sodium on valeraldehyde. As early as 1864 (the beginning of the work on the condensation of aldehydes) Borodin showed that by the action of sodium on valeraldehyde the aldehydate of the metal was not obtained, but rather valeric acid, amyl alcohol, "isocapric" alcohol, $C_{10}H_{22}O$, and a condensation product, $C_{10}H_{18}O$. Subsequently, he found that in addition the acid $C_{10}H_{18}O_2$ was formed and some amount of another condensation product, $C_{20}H_{38}O_3$.[40]

All the products that he described represented the final result of the action of metallic sodium on valeraldehyde. If the reaction is not carried to completion, entirely different intermediate products are obtained. The final products of the reaction are formed spontaneously from these, in fact, and not from valeraldehyde. The initial action of sodium on valeraldehyde consists of the condensation of the latter with elimination of water and formation of sodium hydroxide. Further products result from the action of the caustic alkali and nascent hydrogen on the condensation products. In such a manner the course of the reaction of sodium with valeraldehyde is explained.

Intermediate products of this reaction are:

1. A new polymer [dimer][40] of valeraldehyde, a colorless, thick, viscous liquid, which is insoluble in water, less dense than water, does not react with sodium hydrogensulfite, and does not distill without decomposition. Upon heating it changes again into valeraldehyde. This compound was obtained in pure form by the action of solid potassium hydroxide on valeraldehyde at 0°.
2. A product of further condensation with the elimination of water has the composition $C_{20}H_{38}O_3$. It is an oily liquid with a faint odor and a very bitter taste. It distills without decomposition and boils at about 260–290°[41]. Its specific gravity is 0.895 to 0.900, and it does not react with sodium hydrogensulfite. This compound, on heating with alkalis, is decomposed, giving valeric acid, amyl alcohol, and a little valeraldehyde.
3. A product of the dimerization (with elimination of water) has the composition $C_{10}H_{18}O$ and is an oily liquid, which is easily oxidized to the acid $C_{10}H_{18}O_2$.
4. The product $C_{20}H_{38}O_3$ was formed after removal of water from the polymer [dimer] of valeraldehyde referred to in (1). Valeric acid and amyl alcohol were obtained after decomposition of this compound under the influence of alkali; the alcohol $C_{10}H_{22}O$ was obtained as a result of the attack of nascent hydrogen on the aldehyde $C_{10}H_{18}O$.
5. The compound $C_{10}H_{18}O$ was obtained by means of condensation without elimination of water. The acid $C_{10}H_{18}O_2$, analogous to "isocapric" acid, which was described earlier by Borodin, was obtained after oxidation by the action of sodium hydroxide on heating.

Investigating in detail the action on valeraldehyde of alkalis, hydrochloric acid, the anhydride of phosphoric acid, phosphorus pentachloride, zinc chloride, and other substances, Borodin found that with all these reagents the same products of condensation (with the elimination of water) are obtained.

As we see, in his detailed study of the condensation reaction of aldehydes, Borodin used chemical compounds which later were widely used in the practice of organic chemistry as "condensing" agents.

Borodin's second report [*B 24*] concerned the action of sodium and caustic alkalis on enanthaldehyde[42] at high temperature. It was shown that the behavior of enanthaldehyde is quite analogous to the behavior of valeraldehyde, the only difference being that the elimination of water occurs here much more easily. By the action of solid potassium hydroxide in the cold, enanthaldehyde forms two polymers, one, a hard crystalline substance and the other, an oily liquid, which upon distillation produces enanthaldehyde and subsequent products of the condensation with elimination of water.

Borodin gave an account of the most interesting and important material in his third report [*B 25*]. Studying the reaction of the condensation of acetaldehyde, Borodin discovered a new chemical compound, well known in organic chemistry under the name of aldol.[43]

Aldol represents the hydroxyaldehyde which is formed from acetaldehyde under the influence of catalysts (hydrochloric acid, $ZnCl_2$, carbonates, acetates), particularly under the influence of caustic alkalis, according to the following reaction:

$$2\,CH_3CHO \rightarrow CH_3CH(OH)CH_2CHO\,.^{44}$$

On heating, aldol loses water and forms crotonaldehyde:

$$CH_3CH(OH)CH_2CHO \rightarrow CH_3CH{=}CHCHO + H_2O\,.$$

Resins, formed under certain conditions from acetaldehyde, were obtained at the expense of simple bimolecular condensation, of which aldol appeared to be the main intermediate product.

On May 4, 1872, at a meeting of the Chemical Society, Borodin reported on the physico-chemical properties and the conditions for the formation of the compound which he had recently discovered[45] [*B 25*]. Limiting the report to an authoritative substantiation, Borodin foreshadowed further investigations of the condensation of aldehydes. At this time (1872) Wurtz's[46] article[47] appeared in which a chemical compound was described, identical with the substance obtained by Borodin.

Learning about Wurtz's report, Borodin produced only a short article about his work in the Berichte of the German Chemical Society[48] [*B 26*] and did not develop further research in this field. When Borodin was asked why he let Wurtz have the research on aldol, he answered with bitterness:

> My laboratory barely exists on those means which are available for its direction. I do not have even one assistant, whereas Wurtz has enormous means and works with twenty hands, thanks to the fact that he is not ashamed to overload his laboratory assistants with dirty work[49].

These words give a clear characterization of working conditions of the Russian chemist-experimenter in Czarist Russia.

Yet during Borodin's lifetime his research on the condensation of aldehydes was highly rated by Russian chemists. However, only in our own time has the enormous significance of Borodin's research in this field become apparent. By his work on aldol Borodin laid the foundation of the aldol condensation, which is widely used at the present time. It is sufficient to remember that aldol resins, obtained by means of the aldol condensation, have found their use in various branches of industry. Substituting for natural shellac (the clear form of natural Indian resin under the name of gum lac), aldol resins are used in the electrotechnical industry (for their adhesive and gluelike qualities), in the varnish industry for the preparation of alcoholic and insulation lacquers, in the furniture industry for various polishes, etc.

However, this same reaction of the condensation of organic compounds (in the present case, of aldehydes), which was worked out for the first time by Borodin, has an even greater significance for science and technology. It is well known that on the basis of the condensation reaction the most valuable plastics are obtained, without which, as without synthetic rubber, it is impossible to conceive of modern technology.

Even this incomplete list of the uses for the products of the aldol condensation shows how important Borodin's theoretical researches were.

The flourishing of chemical science in Russia during the second half of the nineteenth century attracted the attention of chemists all over the world. During this period many outstanding works by Russian scientists, appearing in Russia, were reprinted in foreign journals. Borodin's works can serve as a good example of this. As soon as they appeared in print, they were immediately published [50] in the leading foreign journals. As, for example, when Borodin's work, Issledovaniia khimicheskogo stroeniia gidrobenzamida i amarina (Investigations of the Chemical Structure of Hydrobenzamide[51] and Amarine) [*B1*] was published in the Bulletin of the St. Petersburg Academy of Sciences in 1858, in the same year a summary of this investigation of Borodin's appeared in the French journals, Répertoire de chimie pure [and] Bulletin de la Société Chimique de Paris, as well as in the German journals, Annalen der Chemie und Pharmazie von J. Liebig and Jahresbericht über die Fortschritte der Chemie. Other chemical works of Borodin's were published in full in these journals.

For example, Borodin's chemical investigations: O deĭstvii iodistogo etila na benzoilanilid (On the Action of Ethyl Iodide on Benzoylanilide) [*B3*] (1858), O deĭstvii iodistogo etila na benzidin (On the Action of Ethyl Iodide on Benzidine) [*B5*], O bromovalerianovoĭ i bromomaslianoĭ kislote (On Bromovaleric and Bromobutyric Acids) [*B7*], O nekotorykh proizvodnykh benzidina (On Several Derivatives of Benzidine) [*B6*] (1860), O deĭstvii slozhnykh efirov na tsinketil (On the Action of Esters on Zinc Ethyl) [*B8*] (1861), K istorii benzila (On the History of Benzil) [*B11*], O deĭstvii tsinketila na khloriodoform (On the Action of Zinc Ethyl on Chloroiodoform [*B12*], O ftoristom benzoile, k istorii ftoristykh soedineniĭ (On Benzoyl Fluoride and the History of Fluorine Compounds) [*B10*], O deĭstvii natrium-amilata na benzil (On the Action of Sodium Amylate on Benzil) [*B11*] (1862), O deĭstvii natriia na valerianovyĭ al'degid (On the Action of Sodium on Valeraldehyde) [*B14*] (1864) were summarized in the year of their publication in such foreign journals as Annalen der Chemie und Pharmazie, Jahresbericht über die Fortschritte der Chemie, Zeitschrift für Chemie und Pharmazie, and others.

It is well known that from the moment of the Russian Chemical Society's founding and the beginning of the publication of the remarkable Zhurnal Russkogo khimicheskogo obshchestva (Journal of the Russian Chemical Society), correspondents from foreign scientific chemical societies residing in St. Petersburg immediately reported the outstanding investigations of Russian chemists abroad.

Beginning in 1869, summaries of Borodin's works were published in foreign journals after reports of them had appeared in the Zhurnal Russkogo khimischeskogo obshchestva (Journal of the Russian Chemical Society). Such investigations of Borodin's as O produktakh deĭstviia parov broma na serebrianye soli kislot maslianoĭ i valerianovoĭ (On the Products of the Action of Bromine Vapors on Silver Salts of Butyric and Valeric Acids) [*B18*], Produkty uplotneniia al'degidov (Condensation Products of Aldehydes) [*B20*] (1869), O proizvodnykh odnogo izokaprinovogo riada (On Derivatives of One of the Isocapric Series) [*B21*] (1870), O deĭstvii natriia na valeral'degid (On the Action of Sodium on Valeraldehyde) [*B23*], O deĭstvii natriia na enantol (On the Action of Sodium on Enanthol[52]) [*B24*], O poluchenii produkta uplotneniia obyknovennogo al'degida (On Obtaining the Product of the Condensation of Common Aldehyde) [*B25*] (1872), O novykh proizvodnykh valeral'degida (On New Derivatives of Valeraldehyde) [*B26*], O deĭstvii ammiaka na kuminol (On the Action of Ammonia on Cuminol) [*B30*] (1873), O nitrozoamarine (On Nitrosoamarine) [*B31*] (1875), O novom sposobe opredeleniia azota v moche (On a New Method of Determination of Nitrogen in Urine) [*B32*] (1876), and others were published in all of the foreign journals mentioned above.

All this shows us how much foreign countries were interested in Borodin's works.

Footnotes to Chapter 8

[1] See Appendix VII. — A.

[2] Nikolaĭ Aleksandrovich Menshutkin (1842–1907) was a well-known Russian chemist and professor at St. Petersburg University from 1869. From that same year until 1900 he was editor of the Chemical Section of the Zhurnal Russkogo fiziko-khimicheskogo obshchestva (Journal of the Russian Physical-Chemical Society). — A. [See Y. I. Soloviev: Nikolay Aleksandrovich Menshutkin. In: DSB, 1974, Vol. 9, pp. 304–305. — T.]

[3] C_2H_2, — T.

[4] Oscar Liebreich (1839–1908) was a German pharmacologist and professor at Berlin. He discovered the soporific chloral hydrate and other substances. — A. [See A. Langgaard: Oscar Liebreich. Berichte der Deutschen Chemischen Gesellschaft. *41*: 4801–4804 (1908). — T.]

[5] Liebreich investigated brain substance and described what he called *protagon*, $C_{116}H_{241}N_4PO_{22}$ (Annalen der Chemie und Pharmacie. *134*: 29–44 (1865); See also Fritz Lieben: Geschichte der physiologischen Chemie, Franz Deuticke, Leipzig and Vienna, 1935, p. 545). Liebreich obtained a decomposition product of protagon as a base, which he called *neurine*. He also obtained neurine, glycerophosphoric acid, and fatty acids by boiling protagon with baryta water (saturated barium hydroxide solution). Adolph Strecker had discovered choline in bile (*χολή*) [Comptes rendus hebdomadaires de l'Académie des Sciences, Paris. *52*: 1268 (1962); Annalen der Chemie und Pharmacie. *123*: 353 (1862)], and W. Dybkowsky showed that Strecker's choline and Liebreich's neurine are identical (Journal für praktische Chemie. *100*: 153 (1867)). Adolf von Baeyer showed that neurine is choline hydroxide (trimethylhydroxyethylammonium hydroxide, $[(CH_3)_3N^+CH_2CH_2OH]OH$) (Annalen der Chemie und Pharmacie. *140*: 306 (1866); *144*: 322 (1867)). — T.

[6] Letter of A. P. Borodin to P. P. Alekseev of December 23, 1868, published here for the first time. — A.

7 The bromo acids (RCH(Br)COOH) and acyl hypobromites (RCH_2COOBr) are structural isomers and would therefore yield identical analytical results. The difficulty of preparing pure salts is due to the lack of a reasonably acidic hydrogen atom in the hypobromite. — T.

8 Gavriil Gavriilovich Gustavson (1842–1908) was a well-known Russian chemist and Professor of Organic and Inorganic Chemistry at the former Petrov, now Timiriazev Agricultural Academy. — A. [Together with N. Ya. Dem'ianov he synthesized allene. He proposed a new method of synthesizing cyclopropane by treating an aqueous-alcoholic solution of 1,3-dibromopropane with zinc dust. He also obtained 1,1-dimethylpropane and synthesized chlorine derivatives of cyclopropane. See GSE, 1975, Vol. 7, p. 495. — T.]

9 Nikolaĭ Yakovlevich Dem'ianov (1861–1938) was an outstanding organic chemist and academician. — A. [See GSE, 1975, Vol. 8, p. 112. — T.]

10 Letter from G. G. Gustavson to N. Ya. Dem'ianov of March 6, 1907, Archives of the Academy of Sciences of the USSR, published here for the first time. — A.

11 Capric acid is *n*-decanoic acid, $CH_3(CH_2)_8COOH$, while isocapric acid is 8-methylnonanoic acid, $(CH_3)_2CH(CH_2)_6COOH$. At the time organic nomenclature had not yet been standardized, and the prefix *iso* was merely used to indicate an isomer of the already known compound. From the reactions described below, it is obvious that the compound that Borodin called isocapric acid is actually 2-*n*-propylheptanoic acid, $CH_3(CH_2)_4CH(CH_2CH_2CH_3)COOH$, and its aldehyde and alcohol are derivatives thereof. — T.

12 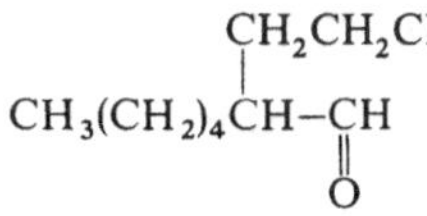 — T.

13 $CH_3(CH_2)_4CH(CH_2CH_2CH_3)CH_2OH$. — T.

14 Possibly, $2\,CH_3(CH_2)_3CHO \xrightarrow{Na} [CH_3(CH_2)_3CH(OH)CH(CH_2CH_2CH_3)CHO]$

$\rightarrow CH_3(CH_2)_3CH{=}C(CH_2CH_2CH_3)\,CHO \xrightarrow{H_2} CH_3(CH_2)_4CH(CH_2CH_2CH_3)CH_2OH$. — T.

15 See Chap. 11. — T.

16 Borodin: Letters, I, p. 150–151. — A.

17 Kekulé, apparently unaware that Borodin had begun work on the condensation of aldehydes in 1864 [*B 14*], began his article, Ueber die Condensation der Aldehyde [Berichte der Deutschen Chemischen Gesellschaft. *3*: 135–137 (1870), received by the journal on February 21, 1870] as follows: "With every chemical work there is the danger, now greater than ever before, that the same experiment will be carried out simultaneously and independently in other places by other chemists. I think it regrettable that many persons consider it suitable to seize upon the investigations of others which are already begun and known by preliminary communications and to continue them further." He then cited his earlier work on the condensation of aldehydes [Condensationsproducte des Aldehyds; — Crotonaldehyd. Ber. *2*: 365–368 (1869)] and remarked, "Recently Riban and Borodine [*sic*] have simultaneously stated that they have begun to work on the same subject, and I will therefore continue the preliminary experiments with valeraldehyde (*Baldrianaldehyd*) no further." In his 1869 paper Kekulé had described the reaction of acetaldehyde with mild reagents (formic acid, potassium acetate, dilute hydrochloric acid, and dilute aqueous zinc chloride) and had acknowledged Charles-Adolphe Wurtz's earlier publication [Justus Liebig's Annalen der Chemie und Pharmacie. *108*: 84–88 (1858)].

Kekulé had been pursuing the question of the bonding in benzene [Ber. *2*: 362–365 (1869)] and, noting that Adolf Lieben [Ann. *106*: 336–338 (1858); Comptes rendus hebdomadaires de l'Académie des Sciences (Paris). *66*: 662 (1858)] and Adolf von Baeyer [*Ann.*, Supplement V, 79–81 (1867)] had condensed three molecules of acetaldehyde in unsuccessful attempts to synthesize benzene, he attempted to condense acetaldehyde with hydrochloric acid and obtained crotonaldehyde. After he became aware of work on the condensation of aldehydes by Borodin [*B 20*] and A. J. Riban [Bulletin de la Société Chimique de France. *13*: 24–25 (1870)], Kekulé responded testily to what he considered an intrusion into an area in which he regarded himself as having already staked a claim, conveniently not rectifying his oversight of Borodin's 1864 paper [*B 14*] and not citing Wurtz and Lieben again.

The episode seems fairly typical of those described as "multiple discoveries" by Robert K. Merton [Singletons and Multiples in Science, Proceedings of the American Philosophical

Society. *105* (5): 470–486 (October, 1961); reprinted in Merton: The Sociology of Science: Theoretical and Empirical Investigations, edited and with an introduction by Norman W. Storer, University of Chicago Press, Chicago and London, 1973; see also Merton's article: Behavior Patterns of Scientists, American Scientist. *57*: 1–23 (1969) and the references therein].

Nevertheless, Kekulé's grasp of the chemistry involved was well ahead of that of his rivals, including Borodin. For a detailed discussion of Borodin's work on the condensation of aldehydes and its relation to work by Kekulé, Wurtz, and others as well as a detailed consideration of Borodin's other research see Ian D. Rae: Borodin's Chemical Research: A Retrospective View on the Centennial of His Death. Ambix, in press. — T.

18 Borodin: Letters, I, p. 202. — A.

19 Butlerov spent 1857–1858 abroad, met Kekulé and Erlenmeyer, and spent about half a year in Paris, where he worked for 2 months in Wurtz's laboratory. This trip resulted in the greatest changes in his work and thought, and when he returned to Kazan in August of 1858, he had changed, in his own words, "from a student to a scholar." — T.

20 For detailed discussions of Butlerov's theory of chemical structure and its priority with respect to Kekulé and other chemists see Henry M. Leicester: Alexander Mikhailovich Butlerov. J. Chem. Educ. *17*: 203–209 (1940); Contributions of Butlerov to the Development of Chemical Theory. Ibid. *36*: 328–329 (1959); G. V. Bykov: Istoriia klassicheskoĭ teorii khimicheskogo stroeniia (History of the Classical Theory of Chemical Structure), Izdatel'stvo Akademii Nauk SSSR, Moscow, 1960; The Origin of the Theory of Chemical Structure. J. Chem. Educ. *39*: 220–224 (1962); G. V. Bykov and L. V. Kaminer: Literatura ob A. M. Butlerova i po istorii klassicheskoĭ teorii khimicheskogo stroeniia (Literature on A. M. Butlerov and the History of the Classical Theory of Chemical Structure), Izdatel'stvo Akademii Nauk SSSR, Moscow, 1962; and G. V. Bykov: Aleksandr Mikhailovich Butlerov. In: DSB, 1970, Vol. 2, pp. 620–625. — T.

21 A. Kekulé: Lehrbuch der organischen Chemie oder der Chemie der Kohlenstoffverbindungen, F. Enke, Erlangen, 1864, Vol. II, Part 2, pp. 244–245. Markovnikov immediately spoke out against Kekulé's historically incorrect contention, but the legend persisted, largely through its promulgation in Carl Schorlemmer's influential: Der Ursprung und die Entwicklung der organischen Chemie, F. Vieweg, Braunschweig, 1889 (The Rise and Development of Organic Chemistry, J. E. Cornish, Manchester, 1879). Recently, John H. Wotiz has cast doubt on another assertion of Kekulé's, viz., his story of the dream leading to his conception of the benzene ring. — T.

22 Until Butlerov's trip there had been a tendency to credit the creation of structural theory to Kekulé, for in 1857–58 Kekulé had stated the idea of the self-linking of carbon atoms which served as one of the preconditions for Butlerov's theory of chemical structure. — T.

23 Lehrbuch der organischen Chemie zur Einführung in das specielle Studium derselben. Aus dem russischen übersetzte: deutsche Ausgabe vom Verfasser revidiert und mit Zusätzen vermehrt, Quandt und Händel, Leipzig, 1868. — T.

24 Central Governmental Archives of the TASSR [Tadzhik Autonomous Socialist Soviet Republic], Stock No. 977, 1867, No. 11135, Sheets 1–3. — A.

25 Borodin: Letters, I, p. 211. — A.

26 In his scientific publications on the condensation of aldehydes [*B 23–25*] Borodin did not reproach Kekulé but merely restated his 1864 priority. — T.

27 See footnote 12, Chap. 2. — T.

28 Borodin: Letters, I, pp. 192–193. — A.

29 Ibid., p. 150. — A.

30 For example,

$$C_6H_5CHO + p\text{-}H_2NC_6H_4NO_2 \rightarrow p\text{-}C_6H_5CH{=}N{-}C_6H_4NO_2 + H_2O. \text{ — T.}$$

31 Filipp Ivanovich Lazarenko was a doctor and a student of Borodin's. Lazarenko's chemical investigations carried out under Borodin's guidance were published in the Zhurnal Russkogo khimicheskogo obshchestva (Journal of the Russian Chemical Society) in 1871–1872. — A.

32 $C_{16}H_{32} + SO_3 \rightarrow C_{16}H_{31}SO_3H$; Reported by Victor von Richter in Ber. *3*: 989 (1870). — T.

33 Borodin: Letters, I, p. 147. — A.

34 Adelaida Nikolaevna Lukanina (née Rykacheva) (1843–?) was a student of Borodin's. In 1872–1874 she studied at Universität Zürich (the University of Zürich). — A. [Paradoxically, although it was not until 1981 that all Swiss cantons except Appenzell had granted women the right

to vote in cantonal elections (they had been allowed to vote in federal elections in 1971), the liberal and progressive Universität Zürich has the distinction of being the first European university to admit women students (1840). During the 1870's the term "Zürcher Studentin" became a famous mark of distinction throughout the continent. — T.]. In 1876 she received the degree of doctor of medicine at Philadelphia. Under Borodin's guidance she carried out chemical investigations which were published in the Zhurnal Russkogo khimicheskogo obshchestva (Journal of the Russian Chemical Society) in 1872–1873. — A.

35 Borodin: Letters, I, pp. 243–244. — A.

36 Reported by Victor von Richter in Ber. *4*: 290 (1871). — T.

37 Pierre [Jacques Antoine] Béchamp [1816–1908] was a French physician and chemist. He was Professor of Chemistry [and Pharmacy] at the University of Montpellier. — A. [, then at Nancy, and finally dean of the Catholic University of Lille. — T.]

38 $2\ C_6H_5COCH(OH)C_6H_5 + ClCOCH_2CH_2COCl$

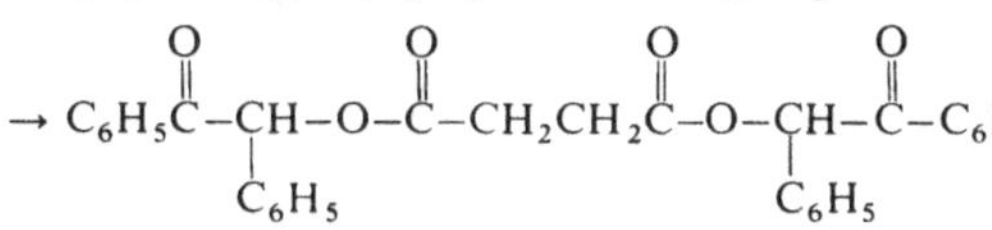

Abstracted by Victor von Richter in *Ber. 5*: 331 (1872) and by T. B. as Succinyl-Benzoin. *J. Chem. Soc. 25*: 1094 (1872). — T.

39 Heinrich [Franz Peter] Limpricht (1827–1909), a German chemist, [was Associate Professor at Giessen (1854) and Professor at Greifswald (1859). The first to adopt the type theory in Germany (1855), he synthesized *iso*-leucine from valeraldehyde and anthracene from benzyl chloride, and he prepared aldehydes by distilling the calcium salts of acids with calcium formate, independently of Piria. — T.].

40 Strictly speaking, this is a condensation product rather than a dimer or polymer. — T.

41 This boiling point seems low for a compound of this molecular formula. — T.

42 Heptanaldehyde or heptanal, $C_6H_{13}CHO$. Borodin was extending his work with the five-carbon compound valeraldehyde (pentanal) to the corresponding seven-carbon aldehyde, enanthaldehyde, which he prepared (A. Borodin: Zhur. Russ. Khim. *1* (1): 31–32 (1869) [*B 18*]) by destructive distillation (pyrolysis) of castor oil (ricinus oil), which consists primarily of ricinolein (the glyceride of ricinoleic acid):

$$\underset{\text{Ricinolein}}{C_6H_{13}CHOHCH_2CH{=}CH(CH_2)_7COOgly} \xrightarrow{\Delta} \underset{\text{Enanthaldehyde}}{C_6H_{13}CHO} + \underset{\text{9-Hendecenoic acid glyceride}}{CH_3CH{=}CH(CH_2)_7COOgly},$$

where gly = glycerol(glycerin). — T.

43 The name was chosen because the compound possesses bith an aldehyde ("ald") and alcohol ("ol") function. — T.

44 Acetaldol, 3-hydroxybutanal, or simply aldol. — T.

45 A. Borodin: Zhur. Russ. Khim. *4* (6): 209 (1872). — T.

46 Charles-Adolphe Wurtz (1817–1884) was a French chemist. — A. [See John Hedley Brooke: Charles-Adolphe Wurtz. In: DSB, 1973, Vol. 14, pp. 529–532. — T.]

47 Compt. rend. *74*: 1361 (1872); *76*: 1165 (1873). — T.

48 [A Borodin: Über einen neuen Abkömmling des Valeraldehyds. — T.] Berichte der deutschen Chemischen Gesellschaft. *6*: 982–985 (1873). — A.

49 M. Goldstein: A. P. Borodin, an article in: [K. K. Arsen'ev and F. F. Petrushevskiĭ, eds.] Entsiklopedicheskiĭ slovar' (Encyclopedic Dictionary), F. A. Brokgauz', I. A. Efron', St. Petersburg, 1893, Vol. 4, p. 440. — A.

50 As summaries or abstracts. — T.

51 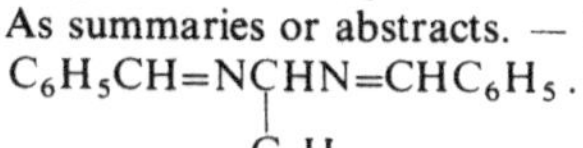 — T.

52 Heptanol or heptyl alcohol, $C_7H_{15}OH$. — T.

Chapter 9

Borodin and Education for Women

At the beginning of the 1870's Borodin played an active and intense role in the organization of higher medical courses for women at the Medical-Surgical Academy.

During the 1860's, in an age of social upsurge, in an awakening of a consciousness in which the flaming articles of the revolutionary democrats Herzen, Belinskiĭ, Chernyshevskiĭ, and Dobroliubov played no small role, leading Russian women displayed a persistent aspiration for higher education. The progressive figures of Russian science came forward in the struggle for the rights of women for higher education and intelligent work. In his article Voprosy zhizni (Problems of Life), written in 1856, the famous Russian surgeon Nikolaĭ Ivanovich Pirogov[1] wrote: "Women must take a place in society, which is all the more responsible for their human dignity and their intellectual faculties"[2]. Pirogov considered that "if a woman receives the proper education and training, then she can establish her scientific, artistic, and social standard of culture as well as a man"[3].

At that time few shared Pirogov's opinion. It is known that in the 1850's, when women began to attend university lectures for the first time, they provoked the ridicule not only of the students but also of some reactionary-minded professors. By all possible measures the Czarist government blocked the Russian women's path to education and in particular prohibited them from attending university lectures. However, the aspiration of Russian women for an education was so great that no power was able to stop it. Combined with the revolutionary movement, the leaders of the female youth of Russia exhibited a persistent desire to study science and to devote themselves to the service of the people.

In Avtobiograficheskie zapiski (Autobiographical Notes) the most prominent Russian physiologist I. M. Sechenov wrote:

> In this year [1863. — A.] entrance to the Medical Academy was closed to women, and both of my women pupils [Nadezhda Prokof'evna — T.] [Suslova and Bokova. — A.], continuing to burn with the desire to live independently steppes.[4] The fact is that at this time the desirability of having women medical doctors for the inhabitants of the Mohammedan steppes has been announced by the authorities of the Orenburg Territory, the reason being the fact that Mohammedan women persistently avoid the care of male physicians. Hearing about this, both young enthusiasts made up their minds to give the authorities a signed statement that they would set off for the steppes if they would only be allowed to study in the Academy[5].

At the end of the 1860's the well-known Russian scientists, the botanists A. N. Beketov[6] and A. S. Famintsyn, the physicist Sergeĭ Alekseevich Usov, the chemists D. I. Mendeleev and A. M. Butlerov, and the physiologist I. M. Sechenov, fervently supported the idea of organizing higher courses for women.

The progressive representatives of Russian science solicited the Minister of Public Education for the organization of higher courses for women, for which they consented to read lectures free of charge. Despite this appeal from the scholars, despite 400 signatures of women who wanted to receive higher education, the Czarist government declined these requests, and instead of the anticipated higher courses or a women's university only public lectures at the university were allowed for women by "Imperial" decree. The Minister of Public Education wrote:

> Although such a solution to the problem does not completely answer the wishes expressed in the writing from the Russian women, in view of the impossibility of other arrangements, they nevertheless have a basis for believing that the majority of women will regard the arrangement of systematic, strictly scientific, public lectures with sympathy[7].

On January 20, 1870 the St. Petersburg public courses were opened. During the first years lectures for the courses were read by the following: in botany, A. N. Beketov and A. S. Famintsyn; zoology, Professors Wagner[8] and Brandt; geology, Professor A. A. Inostrantsev[9]; anatomy, Professor F. V. Ovsiannikov[10]; human physiology, Professor I. M. Sechenov; physics, Professor S. A. Usov; organic and inorganic chemistry, Professors A. M. Butlerov and D. I. Mendeleev. In addition to the lectures, which the majority of the professors read free of charge on holidays and Sundays, practical work was arranged with the auditors, who were able to work in university offices and laboratories during the hours when they were free from the use of students. So in the chemical laboratory, which was assigned for courses by the President of the Russian Technological Society, Kochubeĭ[11], the course auditors studied experimental chemistry under the guidance of the first Russian female chemist, Anna Volkova[12]. Mendeleev often visited the laboratory and followed the course of their work.

The professors brought over their own microscopes, anatomical preparations, and other necessary supplies which the courses did not have in order to perform experimental demonstrations for the lectures.

It was not easy for the young pioneers of women's education. They attained their cherished wish, to be useful to society, at the price of enormous difficulty and deprivation.

> Many of them had to endure the painful struggle with close family members who did not understand and therefore, perhaps, did not sympathize with their aspiration for an education. And for the majority of them there was not even the means for support[13].

Some women, not having the opportunity to attend the St. Petersburg public courses, unofficially[14] attended the Medical-Surgical Academy where they studied in the anatomical theater of the famous anatomist Professor V. Gruber.

One of the auditors related: "We hurried stealthily to the academy to listen to the lectures of Professors Sechenov and Botkin, who made a strong impression on our young minds"[15].

At the beginning of the 1870's the problem of the fight against epidemics, which took an enormous number of human lives, was widely discussed in Russian newspapers and journals. At the same time attention was directed to the high death rate of children, because of the complete absence of medical services, which aroused the population. In connection with this, the problem of training women doctors and midwives was very frequently discussed. These questions received such extensive coverage in the press and were discussed so animatedly in society that under the pressure of public opinion the Czarist government was forced to deal with the fight against epidemics and child mortality even if only by elementary measures. As a result of all this, the government's permission was finally obtained for the organization of courses of women obstetricians at the Medical-Surgical Academy. These were begun in the autumn of 1872.

By permitting training in obstetrics courses for young Russian women the Czarist government hoped to oppose their aspiration to receive a higher education in humanitarian and natural scientific professions. Most of all, the Czarist government was afraid that the spread of education among Russian women would incite the growth of the revolutionary movement. The higher medical course for women was the first establishment of this type in the history of education for women, not only in Russia but also in the entire world.

Mariia Vasil'evna Trubnikova, Borodin, N. V. Stassova[16], M. G. Ermolova[17], and others took an active part in the organization of the medical courses for women. The decision concerning the opening of the courses was not easy to put into practice. In order to achieve any normal performance of teaching, the initiators of the courses had to overcome the passivity of Czarist officials in the solution of the majority of the practical problems and had to pierce the thick wall of conservatism. The Czarist officials opposed the introduction into the program of subjects of general education which could raise the cultural level of the future women physicians. Borodin had to work a great deal in order to provide the courses with a normal amount of knowledge in chemistry. After great efforts he succeeded in opening a chemical laboratory where the auditors of the midwives' courses could conduct practical work. Borodin's chemical laboratory in the Medical-Surgical Academy was the first Russian laboratory which opened its doors to women.

Borodin began reading chemistry for the medical courses for women in the autumn of 1872. All the women students who heard lectures in the theoretical chemistry course also did practical work in the chemical laboratory under Borodin's guidance.

Some of the auditors, wanting to study chemistry seriously, put in extra time in Borodin's laboratory. In this way Adelaida Lukanina carried out two chemical works under Borodin's guidance, of which one, O deĭstvii khloristogo suktsinila na benzoin (On the Action of Succinyl Chloride on Benzoin),[18] was published in the Bulletin of the St. Petersburg Academy of Sciences. This work of Lukanina's, as A. P. Dianin correctly noted, is almost the first scientific work by a woman to appear in publications of the Russian Academy of Sciences.

Borodin eagerly observed the work of the women students. Always helping the needy as much as he could, he tried to organize the work of the graduating women physicians. He arranged musical concerts on behalf of the auditors of the courses and actively took part in the charitable Society for Grant-in-Aid to the Students of the Women's Medical and Pedagogical Courses. Borodin's versatile and energetic support of the courses continued up to 1887.

Occupation with the women's courses distracted Borodin not only from his scientific work but also from his music.

Rimsky-Korsakov wrote,

> Borodin, who had always given but little of his time to music and who often said (when reproached for it) that he loved chemistry and music equally well – began to devote still less time to music than before. Yet it was not science that enticed him. He had become one of the prominent workers in establishing medical courses for women and had begun to participate in various societies for the aid and support of student-youth, especially women. The meetings of these societies, the office of treasurer, which he filled in one of them, the bustling, the solicitations in their behalf, came to take up all of his time. Rarely did I find him in his laboratory, still more rarely at musical composition or at the piano. Usually it turned out either that he had just gone to or was just returned from a meeting; that he had spent all day driving about on those same errands, or else had been writing business letters, or working over his account-books. Add to these his lectures, the various boards and meetings of the academic conference, and it will become clear that there was no time at all left for music[19].

In a letter to L. I. Karmalina[20] Borodin described his life in this period as follows:

> In this long period of time life goes on, however, more badly than well. With the departure of one of our professors of chemistry [Zinin. — A.] part of the work load which the latter carried fell upon me. And this new activity, which required a preliminary organization of educational work, took much time away from me. Besides, our Academy is on trial and is awaiting the verdict of its judges. This exceptional transitional phase is a very bad influence upon the entire Academy and consequently on the position of my department. There is a great deal of trouble, anxiety, and necessary and unnecessary fuss. Add to this the troubles of slow progress and financial embarrassment in the department. All this is not conducive to a good frame of mind and leaves very little spare time for my favorite occupations. Domestic affairs also do not fare brilliantly – my poor wife is always ill and this year more than the previous one. One thing that makes me somewhat happy is the business of the women's courses, which, although they take up much of my time, give a moral satisfaction, perfectly living up to all expectations. Because of educational and scholarly activities, all sorts of commissions, committees, meetings, etc., etc., there is almost no spare time left for music. Only in snatches do I try to find a minute in order to look at something new, to listen to others, etc. I myself am almost unable to work in the musical field. If there is sometimes the physical spare time, then one does not achieve the moral spare time, the composure which is necessary in order to tune up musically. The head is not ready for it[21].

Footnotes to Chapter 9

[1] See footnote 2, Chap. 2. — T.

[2] E[lena] I[osipovna] Likhacheva: Materialy dlia istorii vysshego zhenskogo obrazovaniia v Rossii 1856–1880 (Materials for the History of Higher Education for Women in Russia 1856–1880) [hereafter abbreviated as Materials, Tipografiia M. M. Stasiulevicha], St. Petersburg, [1899–]1901, p. 2. — A.

[3] Ibid., p. 3. — A.

[4] Steppes are the dry, often treeless grasslands of central Eurasia. The Kirghiz Soviet Socialist Republic, since 1936 a constituent republic of the USSR, borders the Kazakh SSR on the north and northwest, the Uzbek SSR on the southwest, the Tadzhik SSR on the south, and China on the southeast. Incidentally, one of Borodin's most popular orchestral pieces is the symphonic poem: On the Steppes of Central Asia. — T.

[5] I. M. Sechenov: Autobiographical Notes, pp. 132–133. — A.

[6] Andreĭ Nikolaevich Beketov (1825–1902) was a great Russian botanist and professor at St. Petersburg University from 1861 to 1897. From 1873 to 1883 he was rector of the university. He organized the first congresses of naturalists and took an active part in the organization of St. Petersburg public lectures for women. — A. [See GSE, 1973, Vol. 3, p. 105. — T.]

[7] E. I. Likhacheva: Materials, 1901, p. 505. — A.

[8] Nikolaĭ Petrovich Vagner (Wagner) (1820–1907) was a zoologist and writer. He was Professor of Zoology at St. Petersburg University. He established the fact of pedogenesis [the formation and development of soil] for the first time. — A. [See GSE, 1974, Vol. 4, pp. 465–466. — T.]

[9] Aleksandr Aleksandrovich Inostrantsev (1843–1919) was a well-known Russian geologist. From 1873 he was Professor of Geology at St. Petersburg University. — A. [See GSE, 1976, Vol. 10, p. 295. — T.]

[10] Filipp Vasil'evich Ovsiannikov (1827–1906) was a well-known physiologist and histologist. He was an academician from 1864. — A.

[11] Pëtr Arkad'evich Kochubeĭ (1825–1892), a chemist, was one of the founders of the Russian Technological Society in 1866 and of its technical-chemical laboratory. — A.

[12] Anna Fedorovna Volkova, one of the first Russian female chemists, studied chemistry under the guidance of Professor A. N. Engelhardt, in whose laboratory at the Forestry Institute she conducted chemical researches. Volkova's works appeared in the Zhurnal Russkogo khimicheskogo obshchestva (Journal of the Russian Chemical Society) from 1870 to 1873. — A. [According to GSE, 1974, Vol. 5, p. 574, she was "the first woman in the world to publish studies on chemistry." — T.]

[13] E. I. Likhacheva: Materials, p. 540. — A.

[14] On May 11, 1864, by a decree of the Minister of War, women were prohibited from attending the Medical-Surgical Academy. — A.

[15] E. I. Likhacheva: Materials, p. 539. — A.

[16] Nadezhda Vasil'evna Stassova (1822–1895), a prominent public figure, was V. V. Stassov's sister. — A. [See GSE, 1975, Vol. 9, pp. 534, 535; 1980, Vol. 24, p. 481. — T.]

[17] Mariia Gavrilovna Ermolova was an energetic figure in the organization of higher education for women in Russia. With the initiation of the Doctor's Courses for women in the autumn of 1881 she was made honorary inspector. — A.

[18] See p. 126. — T.

[19] N. A. Rimsky-Korsakov: My Musical Life, pp. 163–164. — A. [English version, pp. 162–163. — T.]

[20] Liubov' Ivanovna Karmalina (died in 1903), a singer, was a friend of the Borodin family and a student of Glinka's and Dargomyzhskiĭ's. — A.

[21] V. V. Stassov: Letters, p. 4. — A.

Chapter 10

A. P. Borodin – Scientific Leader

On August 20, 1873 the IVth Congress of Russian Naturalists and Physicians opened at Kazan. Borodin was invited to the Congress from the Medical-Surgical Academy and was elected a member of the Congress' administrative committee. After a general organizational meeting, several sections, which met separately, were created. At the first meeting of the chemical section Butlerov presided, at the second, Borodin. Borodin wrote to his wife,

> In our chemical section there were many interesting reports, and among them, I say this without boasting, mine were among the most notable. The quality and number of them (7 items) greatly impressed all members of the section and advanced our laboratory in the opinion of the chemists and even the non-chemists[1].

At the first meeting of the combined sections of chemistry, mineralogy, geology, and paleontology on August 22, 1873 Borodin gave several reports. The first of them was devoted to the work of his student Lobanov, O iodoproizvodnykh fenola (On Iodine Derivatives of Phenol), which had been carried out in the chemical laboratory of the Medical-Surgical Academy. The isomers of monoiodophenol (*meta*-iodophenol, *ortho*-iodophenol, and *para*-iodophenol)[2] were studied in detail by Lobanov.[3] The second report concerned Medical Doctor M. I. Shalfeev's[4] work, Ob okislenii rutovogo masla i o nekotorykh proizvodnykh pelargonovoĭ kisloty (On the Oxidation of Rue Oil and on Some Derivatives of Pelargonic Acid),[5] which confirmed the facts that rue oil contains a ketone and that, upon oxidation, in accord with Popov's[6] general principle concerning the oxidation of ketones, it gives pelargonic and acetic acids.[7] In addition to the already well-known derivatives of pelargonic acid, Shalfeev, under Borodin's guidance, isolated the amide of pelargonic acid [$C_8H_{17}CONH_2$] for the first time.[8] Then Borodin gave an account of the work of Aleksandr Pavlovich Dianin, a student of the Medical-Surgical Academy, O deĭstvii soleĭ okisi zheleza na naftoly (On the Action of Ferric Salts on Naphthols).[9] Dianin showed that cautious action of ferric salts in aqueous solution on *alpha*- and *beta*-naphthol oxidizes them according to the equation:

$$2\,C_{10}H_8O \rightarrow C_{20}H_{14}O_2 + H_2\,.$$ [10]

At the meeting of the Chemical Section on August 23rd Borodin reported P. G. Golubev's[11] work, O vosstanovlenii nitrobenzila olovom v solianoĭ kislote ili tsinkom v uksusnoĭ kislote (On the Reduction of Nitrobenzoin by Tin in Hydrochloric Acid or Zinc in Acetic Acid). Golubev showed that in all of these cases nitrobenzoin is converted to aminodesoxybenzoin.[12]

At this same meeting Borodin also reported on his own works, O novom produkte uplotneniia valeral'degida (On a New Product of the Condensation of Valeraldehyde) [*B 26*] and Ob ammiachnykh proizvodnykh kuminovogo al'degida (On Ammonium Derivatives of Cumaldehyde) [p-$(CH_3)_2CHC_6H_4CHO$] [*B 30*]. In the first work he found that upon prolonged standing in a very dilute solution of sodium carbonate the polymer of valeraldehyde, $C_{10}H_{20}O_2$, slowly changed from viscous and turbid to fluid and transparent, and small needle-shaped crystals were formed. Analysis of the crystals showed that they could be considered as the hydrate of the polymer [dimer], $(C_{10}H_{20}O_2)_2 \cdot H_2O$, which, upon heating in a closed container, decomposed to valeraldehyde, water, an aldehyde of the composition $C_{10}H_{18}O$, and the condensation product, $C_{20}H_{38}O_3$.

On the basis of his previous research on ammonium derivatives of cumic aldehyde, which showed that in hydrobenzamide all the hydrogen belongs to the hydrocarbon radical, but that in amarine part of the hydrogen is linked in the amide residue[13], Borodin stated the theoretical considerations concerning the conversion of hydramides into isomeric amines [*B 28*]. Being occupied with the study of these derivatives, Borodin incidentally obtained the hydramide and the isomeric amine of the cumic series, about which there had been only negative data in the literature until his research. He also obtained the sulfate, nitrate, and oxalate salts of the amine.

At the same time Borodin presented a report on the as yet unfinished research on ammonium derivatives of salicylaldehyde and his attempts at their isolation in order to elucidate the structure of amines isomeric with hydramines of hydrocarbon radicals from amarine and its analogs.

Borodin's last report [*B 29*] concerned succinyldibenzoin, $(C_{14}H_{11}O_2)_2C_4H_4O_2$, which has already been discussed. In this report Borodin (together with Lukanina) showed that upon the action of succinyl chloride on benzoin, succinyldibenzoin is obtained, despite the German chemist Limpricht's statement.

All these works of Borodin and his students had great significance in explaining the structure of complex organic compounds, which often remained insufficiently elucidated. Borodin deserves the highly important scientific credit for determining the composition of the products of the condensation of valeric and other aldehydes, succinyldibenzoin, $(C_{14}H_{11}O_2)_2C_4H_4O_2$; amarine,

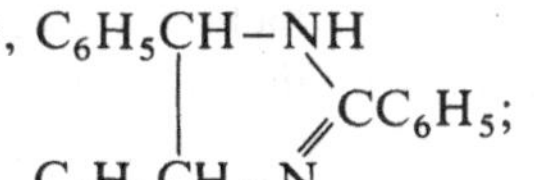

and other compounds.

At the Congress Borodin enjoyed great popularity and was surrounded by universal attention. He wrote to his wife, "The reports and oral debates greatly raised my stock on the part of knowledge. Butlerov, Zaĭtsev[14], and others did not withdraw from me"[15].

At the banquet arranged in honor of the opening of the Congress, Borodin, as the representative of the women's courses, made a speech. He toasted the prosperity of special education for women. As he wrote to his wife,

> an uproar was raised, and they gave me a noisy ovation. Afterwards [news of] all this rapidly spread around Kazan . . . One of the Kazan women sent a message to me through Butlerov saying that she sincerely respects and vigorously thanks me for my sympathy toward women's affairs . . . In general the Congress was as successful as possible for me[16].

Returning from Kazan, Borodin again began work with amarine with the research, of which (in 1858) he had begun his scientific career. During 1873 and 1874 he gave several reports in the Russian Chemical Society on chemical investigations carried out under his guidance in the laboratory of the Medical-Surgical Academy.

On December 13, 1873, at a meeting of the Russian Chemical Society, Borodin reported Doctor Lazarenko's work, O tsetene (On Cetene), and the research of his student A. P. Dianin, O reaktsii dinaftola s khloristym benzoilom (On the Reaction of Dinaphthol with Benzoyl Chloride), and on February 7, 1874 he reported Golubev's work, O dvunitroazobenzoĭnoĭ kislote (On Dinitroazobenzoic Acid),[17] and Dianin's further investigations on the products of the oxidation of *alpha*-naphthol.[18]

On May 2, 1874, at a meeting of the Chemical Society, Borodin gave a report on behalf of a student of the Medical-Surgical Academy, M. Yu. Goldstein[19], about the fact that upon careful oxidation of volatile [*ortho* — T.] nitrophenol by potassium permanganate the reaction proceeds analogously to the reaction of the oxidation of naphthol, that is, dinitrodiphenol [dihydroxydinitrobiphenyl. — T.][20] is obtained from two molecules of nitrophenol.

On October 3, 1874, at a meeting of the Chemical Society, Borodin reported on Golubev's behalf that several compounds are formed by the action of ethyl iodide on silver azobenzene-3,3′-dicarboxylate. For one of these Golubev determined the composition of the diethyl ester, $C_{14}H_{18}(C_2H_5)_2N_2O_4$, and the barium and silver salts of its parent acid were both obtained.[21]

On December 5th of the same year Borodin again came forward at a meeting of the Chemical Society and reported that Dianin, in the chemical laboratory of the
composition of the diethyl ester, $C_{14}H_{18}(C_2H_5)_2N_2O_4$, and the barium and silver salts
Medical-Surgical Academy, had obtained dithymol $\begin{matrix}C_{10}H_{12}OH\\ | \\ C_{10}H_{12}OH\end{matrix}$ [22] by the action of ferric chloride on thymol[23] and that by the action of benzoyl chloride he obtained dibenzoyldithymol.[24] He also reported Dianin's work, Ob oksidinaftaline (On Oxydinaphthalene),[25] which was obtained by distillation of dinaphthol with phosphoric anhydride (P_2O_5). Dianin found that the product of the oxidation of *beta*-naphthol is dinaphthol, which with benzoyl chloride gives dibenzoylnaphthol.

On December 4, 1875 Borodin reported at a meeting of the Chemical Society on Dobroslavin's behalf the work, Deĭstvie sliuny na razlichnye vidy krakhmala (The Action of Saliva on Various Forms of Starch) and also on Goldstein's behalf his research on obtaining a dibenzoyl derivative by the action of benzoyl chloride on dinitrodiphenol. The main product of the action of fuming nitric acid on dinitrodiphenol is nitrobenzene.[26]

On October 2, 1875, at a meeting of the Chemical Society, Borodin reported his work with nitrosoamarine, $C_{21}H_{17}(NO)N_2$, which was obtained by the action of potassium nitrite on solutions of amarine salts in the presence of acetic acid

[*B 31*]. In order to learn if amarine is a primary or secondary amine derivative, Borodin undertook the study of the products obtained by the action of salts of nitrous acid on amarine. Borodin reasoned that if amarine is a primary amine derivative, then the formation of diazo compounds or products of their subsequent conversion ought to be expected. If amarine is a secondary amine derivative, then the formation of a nitroso compound, in which the nitroso group would take the place of the hydrogen of the imide group, ought to be expected. Also, in this case the nitroso group, being weakly retained by the nitrogen atom, would have the ability to be easily evolved in the form of an oxide of nitrogen and also to be substituted by hydrogen, and in both cases amarine would be reformed. His research led Borodin to the formation of such a nitroso compound and therefore also to the conclusion that amarine is a secondary amine derivative. Borodin described in detail the physical and chemical properties of nitrosoamarine. He found that acids convert nitrosoamarine to amarine and that with phenol nitrosoamarine yields a dark blue coloration, which diappears with zinc dust. He also showed that the reaction for the formation of nitrosoamarine can serve as a qualitative reaction for salts of amarine.

In 1874 the chemical laboratory of the Medical-Surgical Academy came completely under Borodin's leadership, since Zinin left the academy. After Zinin's departure, Professor N. V. Sokolov[27] read inorganic chemistry for students of the first course. In addition to this he also read the section on physiological chemistry for the second course. Until Zinin's retirement Borodin, beginning in 1862, read the first and second courses on both organic and inorganic chemistry without the practical work. In 1874 he also began to direct the practical work of the students in analytical chemistry.[28] His student A. P. Dianin was his assistant.

Borodin's teacher, Zinin, although an excellent experimenter and lecturer, was, however, overloaded with diverse and numerous duties and was not able to supervise personally the practical work of all the students, whose number reached 400. He had few assistants. Borodin, who fervently preached the necessity for serious scientific education in Russia, decided to organize the work of the students properly. Dianin wrote,

> Recognizing the enormous importance and the essential official role of chemistry in addition to its independent significance, Aleksandr Porfir'evich long ago already had the right idea of organizing the work of the students and medical doctors in the laboratory.[29]

Carrying out this task was not easy for Borodin. Meager means for the laboratory, a shortage of reagents, and the lack of assistants hindered the growth of the project.

In order to give an opportunity of carrying out practical work to 300–400 male and female students Borodin secured the assistance of all the chemists working in the laboratory. Dianin reported that

> few, even of the persons closely associated with Aleksandr Porfir'evich, know what an amount of time, energy, trouble, and even personal expense this project cost him. Once, with his own personal means, he even maintained a private assistant and a spare servant at the laboratory. The first year (1874) produced a

> brilliant result . . . In the following year the work in the laboratory progressed even more successfully; in general, it was clear that the work had become firmly established. The organization of these works drew great praise for A. P. before the Academy.[30]

Borodin's scientific-instructive and pedagogical activity, to which he gave much time and energy, was underrated by his musical friends. Meanwhile, Borodin, training a new cadre of skilled specialists, performed a most important patriotic act. The first Russian women medical doctors, students of Borodin's, selflessly struggled against epidemics of smallpox, typhus, and scarlet fever in remote corners of Russia.

Thanks to such fervent advocates of special chemical education in Russia as A. A. Voskresenskiĭ, N. N. Zinin, A. P. Borodin, N. N. Sokolov,[31] and others, numerous Russian chemical schools were created with their own unique trends. A remarkable galaxy of Russian chemists, an honor to the names of their teachers by discoveries of paramount significance, advanced Russian chemistry to one of the first places in the world.

The widespread popularization of the importance of the natural sciences, which was carried out by the enlighteners and revolutionary democrats, and the scientific achievements of the prominent Russian naturalist-materialists resulted in a change of opinion in Russian society with regard to the natural sciences. In his article Probuzhdenie estestvoznaniia v tret'eĭ chetverti XIX veka (The Awakening of Natural Science in the Third Quarter of the 19th Century), the famous Russian biologist Klimentiĭ Arkad'evich Timiriazev wrote:

> I remember well that when my older brother began to study chemistry, it evoked the bewilderment of the entire family – a family, I should say, which stood considerably higher than their surroundings, in general, and in political attitude, in particular. "For what is chemistry to him", they said. "Is he preparing himself to become a druggist? As far as that goes he should begin to study medicine. Perhaps a second Pirogov[32] will emerge with him." When, after some 5–6 years, I began to study botany, it already did not disturb anyone.[33]

Lectures and practical work for a great number of students in the academy and the women's courses required almost all of Borodin's time. Therefore in the period from 1872 to 1876 Borodin was not able to begin any experimental investigations that demanded continuing and prolonged work in the laboratory. Only in 1876, at the request of his fellow medical doctors, who turned to him for advice, did Borodin study the question of the quantitative determination of nitrogen associated with the study of the conversion of nitrogen in the animal organism. Studying this question, Borodin worked out a new method and proposed his apparatus for the determination of nitrogen. Borodin's method is still widely used by doctors and biologists studying nitrogen exchange in the organism[34].

Borodin's method for the determination of nitrogen (urea) consists of the following: in the apparatus, consisting of a Mohr burette into which another burette is introduced in an inverted position, the urea is acted upon by brominated alkali (a solution of $NaOH + Br_2$), whereupon water, nitrogen, and carbon dioxide are evolved[35]. The carbon dioxide is completely absorbed by the alkali, and the nitrogen content of the urea is calculated from the volume of free nitrogen evolved. The method is sufficiently precise to be used widely for the quantitative determination of urea in urine.

In a letter to A. N. Shabanova, who was occupied with biochemical analysis, Borodin indicated that

> the determination of urea by sodium hypobromite yields a small negative error. The determination by mercuric nitrate yields a much larger positive error . . . According to that method, which you used, my error was not more than 0.5%, and calculating for the urine, i.e., for a tenth (a tenfold diluted solution of it was used by me for analysis), it constitutes 0.05%. It goes without saying that the error is insignificant (sometimes 0.001% and less) if the measurement of the gas is made not after 5 minutes, as is usual, but after 3–5 hours[36].

Fig. 11. Aleksandr Porfir'evich Borodin with his students—graduating physicians of 1878, at the Medical-Surgical Academy. Seated (left to right): D. G. Nikol'skiĭ, A. P. Borodin, P. F. Petermak. Standing (left to right): A. P. Dianin, I. A. Al'bitskiĭ (April–May 1878)

On January 8, 1876, at a meeting of the Chemical Society, Borodin, on Shalfeev's behalf, reported the investigation of cerotinic acid[37] from beeswax.[38] By detailed analysis of the acid of composition $C_{27}H_{54}O_2$, Shalfeev found that it represented a mixture, from which an acid, corresponding to the composition $C_{34}H_{68}O_2$, was isolated.

In 1877, at one of the meetings of the Chemical Society, Borodin reported, on Golubev's behalf, that by the action of fuming nitric acid on desoxybenzoin[39] two nitro compounds are formed: mononitrodesoxybenzoin and dinitrodesoxybenzoin.[40] Aminodesoxybenzoin is obtained from the first compound, apparently identical with that obtained by reduction of nitrobenzil.

At the beginning of 1877 the Russian Chemical Society organized an exhibition, the first in Russia, of chemical preparations and apparatus presented by various chemical laboratories of the most important schools of higher education of Russia.

The following exhibits were sent to the exhibition from the chemical laboratory of the Medical-Surgical Academy: (1) various products of the condensation of valeraldehyde; (2) coumarin and its sulfate, nitrate, chloride, and oxalate salts; (3) nitrosoamarine; and (4) apparatus for the determination of urea by Borodin's method.

An exhibition of apparatus and chemical preparations, sent by well-known chemical laboratories of various countries, had opened in London a year earlier. A large quantity of apparatus and chemical preparations was sent to the exhibition from Russia. In particular, collections of apparatus and preparations of the Medical-Surgical Academy were sent. Englishmen and other foreign scientists who organized this exhibition, recognizing the superiority both qualitatively and quantitatively of the Russian exhibits over their own meager collections, either passed over the original Russian chemical preparations and apparatus in silence or failed to mention them.[41] This unusual incident provoked indignation among Russian chemists. Not confining themselves to a written reproach of the indecent conduct of some foreign scholars, in 1877 the members of the Russian Chemical Society organized an exhibition of chemical preparations and apparatus at St. Petersburg, which was a clear indication of the independence and the enormous successes in the growth of Russian chemical science.

On March 19, 1877 The Conference of the academy elected Borodin an Academician of the Medical-Surgical Academy for his performance of first-class research in the fields of organic and physiological chemistry and also for his productive professorial activity.

In 1880 Russian chemical science suffered a severe loss. One of its founders, Nikolaĭ Nikolaevich Zinin, died. Borodin sadly mourned the death of his dearly beloved teacher and friend. On February 9, 1880 at his grave Borodin delivered a remarkable speech devoted to the memory of the "Nestor[42] of Russian Chemistry".[43]

Shortly after Zinin's burial Borodin organized a collection of money for the construction of a memorial to the great Russian chemist. Not restricting himself to signatures among the chemists of St. Petersburg, Borodin turned to the chemists of Kiev, Kazan, Kharkov, and other universities.[44] In 1883 a bust of Zinin, created by the sculptor Il'ia Yakovlevich Gintsburg (1859–1939), was erected on the staircase landing between the first and second floors of the Medical-Surgical Academy.

Borodin's activity during this period was by no means confined to scientific laboratory investigations. He took an active part in the work of the Russian Chemical Society and in various commissions and committees, which rendered assistance to the young scholars and students and which discussed questions of vital importance in connection with the growth of Russian science. Lively and energetic, in the bloom of his creative strength, Borodin was at the center of the wide social movement of the intelligentsia of the 1860–1870's.

Borodin belonged to a group of progressive Russian scientists who continuously fought with the reactionaries of science who were firmly settled in some of the St. Petersburg schools of higher education. These apologists of autocracy and orthodoxy (foreigners, for the most part) were supported by the Czarist government. Reactionary leadership of the Academy of Sciences and also of the Medical-Surgical Academy, expressing the interests of the ruling classes, set up foreigners in opposition to the progressive Russian scientists, these usually proving to be more obedient champions to conservative ideas in science.

The advent of the reaction in the 1880's was an expression of the politics of the Russian autocracy, which entered the struggle with the worker's movement which was emerging.

However, the rising wave of reaction did not weaken the struggle for the foremost native science of the best representatives of the Russian people. Outstanding Russian chemists such as Mendeleev, Butlerov, Markovnikov, Borodin, and others came forward as courageous champions against reaction, against servility toward foreigners, and for the acknowledgment of native science. The struggle of the prominent scholar-patriots against the reactionaries of science was especially strong in the Academy of Sciences. Timiriazev wrote:

> The hostility to the Academy was to a considerable extent only one of the evidences of repressed general bitterness against everything German which was so prevalent in the very somber time of the reign of Nicholas I [1825–1855. — T.] when, beginning with Benkendorf and Dubel't, Dibich and Kleĭnmikhel', Adlerberg and Brok,[45] and ending with the master and owner of every industrial institution or estate – the pressure of the aliens is felt everywhere apart from the general oppression of the system.[46]

During the 1880's the reactionary "party" in the Academy of Sciences so openly suppressed everything foremost in Russian science that "it was time to speak straightforwardly," as the scholars of the Physical-Mathematical Faculty of Moscow University wrote in a letter, "time to call the unworthy, unworthy".[47]

The struggle of the progressive Russian scholar-patriots with the reactionary party in the Academy of Sciences took the form of open conflict when the question of membership in the academy of the famous Russian scholars, Mendeleev, Sechenov, Mechnikov, Beketov, and Kovalevskiĭ came to a standstill. The disgraceful blackballing of the great Russian chemist Mendeleev from election to the Academy of Sciences incited the violent indignation of the entire Russian scientific community. Borodin, coming out against the shameless stratagem of the reactionaries, wrote to Menshutkin that he fully shared the "profound indignation toward the scandalous attitude of the Academy of Sciences to our great Russian scholar".[48]

Defending Russian science and its representatives, Borodin came forward on the side of the progressive professors against the reactionary party of scholars also in the Medical-Surgical Academy, headed by the President of the Academy, Pëtr Aleksandrovich Dubovitskiĭ [(1815–1868)]. The foreign scholars, who occupied many of the chairs in the Medical-Surgical Academy during the 1870's and 1880's, opposed the advancement in science of talented and prominent young Russians in every way possible.

> Naturally, it was necessary to express displeasure and vigorous ressistance to the obvious preference for everything foreign and to the scornful attitude toward the old figures of the Russian school[49].

Struggling with the foreign dominance in the Medical-Surgical Academy, Borodin began in every way possible to attract Russian scholars to the Academy. During the 1870's he proposed Mendeleev's candidacy for the vacant Chair of Physics to the Conference of the Academy. In 1873 he wrote a reference recommending the scientific work of Mendeleev and Aleksandr Grigor'evich Stoletov in connection with the discussion of the candidacies for the Chair of Physics to the Conference of the Medical-Surgical Academy.[50] Despite the fact that Mendeleev and Stoletov would have been effective additions to the Medical-Surgical Academy, they were not elected to the Chair of Physics, and this chair remained vacant for more than 10 years.

A letter is preserved in the archieves of the Academy of Sciences from Borodin to Butlerov which clearly characterizes Borodin's attitude toward the reactionary scholars in the Medical-Surgical Academy:

> Sincerely respected Aleksandr Mikhaĭlovich, please forgive me for delaying the delivery of the manuscript . . . In accordance with our agreement, I included my personal recollections, which characterize N. N. [Zinin. — A.] as an academic figure and as my teacher, not concerning his scholarly works and activity in the Academy of Sciences . . . To my regret I cannot touch upon many other things in order not to tease the geese, of whom there are many in life and who do not object to pecking away at the deceased even after death[51].

However, the severe and gloomy working conditions and the oppressive atmosphere, which were created in the 1870's and 1880's in the Medical-Surgical Academy by its reactionary and bureaucratic leadership, were able neither to suppress Borodin's creative aspirations nor to weaken his will. Borodin, a great patriot, carried the banner of the leading figures of Russian science and culture with honor and dignity.

Footnotes to Chapter 10

1 Pis'ma A. P. Borodina (Letters of A. P. Borodin), Vol. 2 [hereafter abbreviated as Letters, II], Moscow, 1936, pp. 38–39. — A.

2 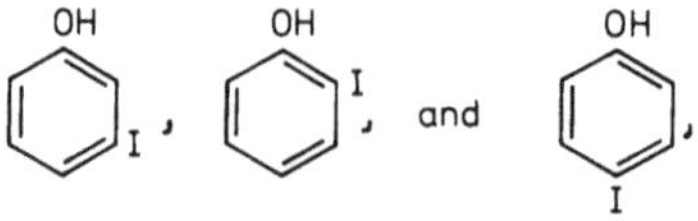

respectively. — T.

3 Abstracted by Victor von Richter in Ber. *6*: 1251–1252 (1873). — T.

4 Mikhail Ivanovich Shalfeev (1845–1910), a student of Borodin's, subsequently became Professor of Physiological Chemistry at Warsaw University. — A.

5 Nonanoic acid, $CH_3(CH_2)_7COOH$. — T.

6 Aleksandr Nikiforovich Popov [ca. 1840–1881], a famous Russian chemist, was a student of Butlerov's at Kazan University. In his master's thesis, Ob okislenii odnoatomnykh ketonov (On the Oxidation of Monatomic Ketones) Popov stated the general rule for the oxidation of

ketones. — A. [See G. V. Bykov: Aleksandr Nikiforivich Popov. In: DSB, 1975, Vol. 11, pp. 92–93 and GSE, 1979, Vol. 20, p. 408. — T.]

7 Popov had found that in the oxidation of ketones the radicals form the following order in their affinity for the former carbonyl group: C_6H_5 and $R_3C > CH_3 > RCH_2 > R_2CH > C_6H_5CH_2$ (where R = an alkyl group), and that in the oxidation of ketones the carbonyl group remains joined to the most stable radical. Popov's rule was used by himself and other chemists to determine the structures of acids, alcohols, and organometallic compounds used for the synthesis of ketones. — T.

8 Abstracted by Victor von Richter in Ber. *6*: 1252 (1873). — T.

9 Ibid. — T.

10 These reactions may be written as follows:

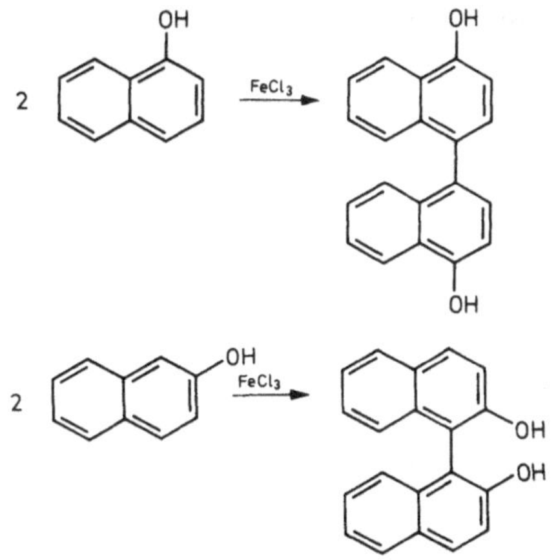

— T.

11 Porfiriĭ Grigor'evich Golubev (1838–1914) was a doctor of medicine and a studnet of Borodin's. From 1875 Golubev was N. V. Sokolov's laboratory assistant. From 1889 to 1891 he was in the Department of Organic Chemistry at the Medical-Surgical Academy (for further biographical details see: Zhurnal russkogo fiziko-khimicheskogo obshchestva (Journal of the Russian Physical-Chemical Society). *47* (1): 1 (1915). — A.

12 Abstracted by Victor von Richter in Ber. *6*: 1252–1253 (1873) and by C. S. (Carl Schorlemmer) as: Reduction of Nitrobenzile by Tin in J. Chem. Soc. *27*: 273 (1874). See also p. 87–88. — T.

13 [A. P. Borodin: Recherches sur la constitution chimique de l'hydrobenzamide et de l'amarine.] Bulletin de l'Académie des Sciences de St. Pétersbourg, [*17*,] (1858) — A. [No. 1, 2, 3, notes, cols. 38–46 [*B 1*], — T.]

14 Aleksandr Mikhaĭlovich Zaĭtsev (1841–1910), a famous Russian organic chemist, was a student of Butlerov's and a professor at Kazan University. — A. [He discovered sulfoxides in 1886. For biographical details see GSE, 1975, Vol. 9, p. 562. — T.]

15 Borodin: Letters, II, p. 39. — A.

16 Ibid., p. 38. — A.

17 2,2'-, 3,3',- or 4,4'-dinitroazobenzenedicarboxÿlic acid, $HOOC(NO_2)C_6H_3N{=}NC_6H_3(NO_2)\cdot COOH$. — T.

18 Reported by A. Kuhlberg in Ber. 7: 487 (1874). — T.

19 Mikhail Yul'evich Goldstein (1853–1905) was a chemist and a student of Borodin's at the Medical-Surgical Academy. In 1877 he received the degree of Doctor of Philosophy at Jena University. Subsequently he was a reader at St. Petersburg University. While in political exile in Arkhangel'sk (Archangel) he was killed at the time of the Black Hundred pogrom in October, 1905. — A. [The Black Hundreds or Soiuz Russkogo naroda (League of the Russian People) was an organization of reactionary, anti-Semitic groups in Russia formed during the 1905 revolution and having the unofficial approval of the government. Composed primarily of landlords, rich peasants, bureaucrats, police officials, and clergymen, who supported orthodoxy, autocracy, and Russian nationalism, they staged raids against various revolutionary groups and pogroms against the Jews. — T.]

[20] Probably

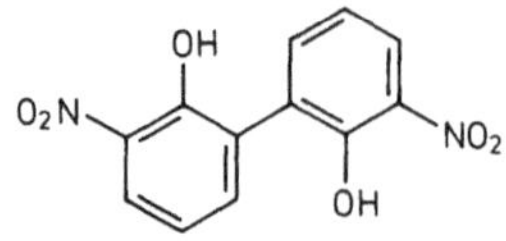

— T.

[21]

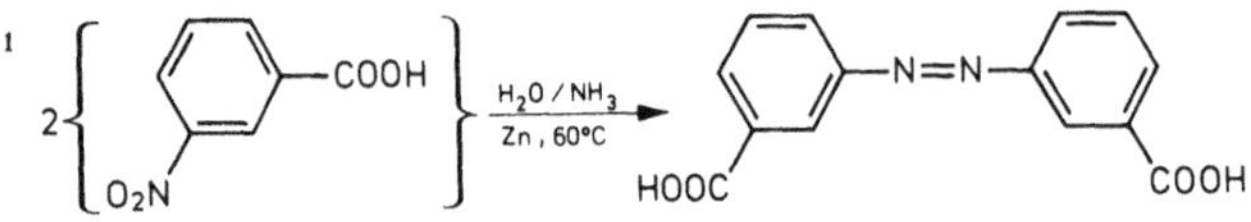

3-Nitrobenzoic acid — Azobenzene-3,3′-dicarboxylic acid

P.G. Golubev: Zh. Russ. khim. *6*: 196 (1874)

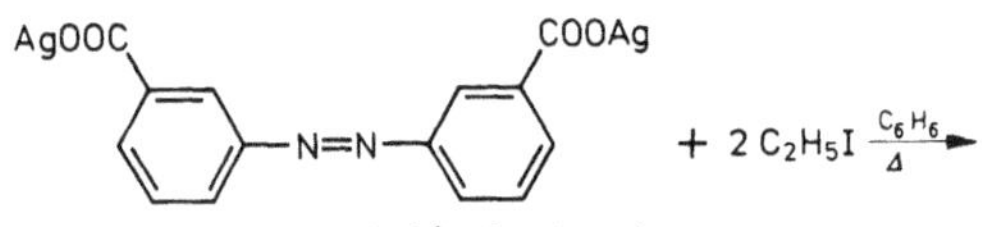

Disilver azobenzene-3,3′-dicarboxylate

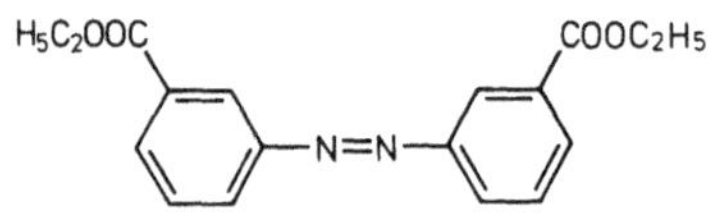

Diethyl azobenzene-3,3′-dicarboxylate

P.G. Golubev: Zh. Russ. khim. *6*: 251 (1874); *16*: 412 (1884). — T.

[22] Probably

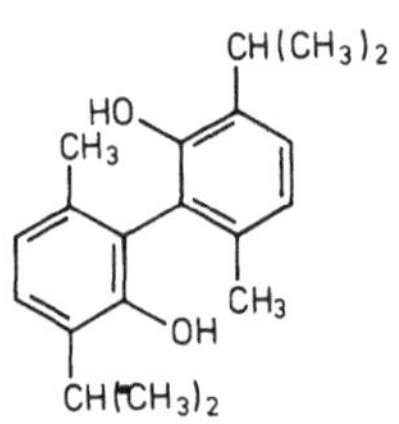

— T.

[23] 2-Hydroxy-1-isopropyl-4-methylbenzene,

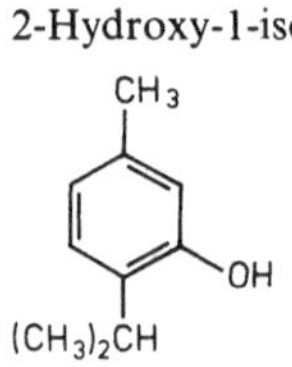

— T.

[24] Reported by A. Kuhlberg in Ber. *8*: 166 (1874). See also A. Kuhlberg: Ber. *7*: 126, 487 (1874). — T.

[25] Reported by A. Kuhlberg in Ber. *8*: 166 (1874). — T.

[26] This seems difficult to believe. — T.

[27] Nikolai Vasil'evich Sokolov (1841–1915) was Professor of Chemistry at the Medical-Surgical Academy (1870–1892). — From the fall of 1874 to 1877 he taught physics for the courses for women medical doctors. — A.

[28] Students made use of the following chemistry books: (1) Osnovy khimii (Foundations of Chemistry) by Mendeleev, (2) Neorganicheskaia khimiia (Inorganic Chemistry) by Richter, (3) Organicheskaia khimiia (Organic Chemistry) by Schorlemmer, (4) Analiticheskaia khimiia (Analytical Chemistry)

by N. A. Menshutkin, and (5) Fiziologicheskaia khimiia (Physiological Chemistry) by Kiune. There were also lectures by the professors. — A.

29 A. P. Dianin: Borodin, p. 373. — A.

30 Ibid. — A.

31 Nikolaĭ Nikolaevich Sokolov (1826–1877) was a famous Russian chemist. Together with A. N. Engelhardt (1828–1893), he founded the Khimicheskiĭ zhurnal (Chemical Journal) (1859 to 1860) and a private chemical laboratory. — A. [See GSE, 1980, Vol. 24, p. 281. — T.]

32 See footnote 2, Chap. 2. — T.

33 K. A. Timiriazev: Awakening, p. 4. — A.

34 This is no longer true today (1986). In clinical urinalysis Borodin's method (Zhur. Khim. Fiz. *8* (5): 145 (1876); *9* (6): 240 (1877); Ber. *9*: 1029 (1876) [*B 33, 34*]) is now known as Ambard's method (L. Ambard: Bull. Soc. Chim. Biol. *2*: 205 (1920); Presse Médicale *31*: 753 (1923)). See also W. Knop: Chemisches Zentralblatt *1860*: 244; Z. anal. Chem. *9*: 225 (1870); *14*: 247 (1875); W. Knop and W. Wolf: Chem. Zentr. *1860*: 257; H. J. H. Fenton: J. Chem. Soc. *33*: 300–303 (1878); *35*: 12–16 (1879); William Foster: ibid. *33*: 470–474 (1878). — T.

35 $3\,NaBrO + CO(NH_2)_2 \rightarrow N_2\uparrow + CO_2\uparrow + 3\,NaBr + 2\,H_2O$. — T.

36 Pis'ma A. P. Borodina (Letters of A. P. Borodin), Vol. 3 [hereafter abbreviated as Letters, III], Moscow, 1949, p. 66. — A.

37 Cerotic acid or hexacosanoic acid, $C_{25}H_{51}COOH$, a fatty acid obtained from beeswax, carnauba wax, or Chinese wax. — T.

38 Reported by A. Kuhlberg in Ber. 9: 298–299 (1876). — T.

39 $C_6H_5CH_2COC_6H_5$. Reported by Georg Wagner in *Ber.* 11: 1939 (1878). — T.

40 Presumably, p-$NO_2C_6H_4CH_2COC_6H_5$ and o,p-$(NO_2)_2C_6H_3CH_2COC_6H_5$, respectively. See Bulletin de la Société Chimique de France [2] *35*: 560 (1881) and Journal of the Chemical Society (London), Abstracts *40*: 813 (1881). — T.

41 In a note in Berichte der Deutschen Chemischen Gesellschaft [Ausstellung wissenschaftlicher Apparate in London. *Ber.* *9*: 1–2 (1876)] on the collection of chemical preparations exhibited at the London exhibition in 1876, "the Russian exhibits and in particular, the exhibits of the chemical laboratory of the Medical-Surgical Academy were not mentioned" (see Zhurnal Russkogo khimicheskogo obshchestva (Journal of the Russian Chemical Society), 1877, part 2, pp. 112–113). — A.

42 In Homer's Iliad the role of Nestor, the aged King of Pylos, is largely to incite the Greek warriors to battle and to tell stories of his early exploits, which contrast with his listeners' experiences, which are shown to be soft and easy. — T.

43 Borodin's speech at Zinin's grave was published in the journal Zdorove'e (Health), 1880, No. 135; see Appendix VIII. — A.

44 See Appendix IX. — A.

45 Aleksandr Khristoforovich Benkendorf (1781 or 1783–1844) was a Russian statesman, count (from 1832), and one of the main champions of the reactionary domestic policy of Czar Nicholas I (See GSE, 1973, Vol. 3, p. 169). Leontiĭ Vasil'evich Dubel't (1792–1862), one of the most reactionary figures of Nicholas I's reign, was Chief of Staff of the Corps of Gendarmes, Director of the Third Section (1839–1856), and member of the Central Board of Censorship and of a secret committee on schismatics (See GSE, 1975, Vol. 8, p. 428). Ivan Ivanovich Dibich-Zabalkanskiĭ (1785–1831), the son of a Prussian officer who switched to the Russian service in 1798, was a Russian field marshal (from 1829) and count (from 1827), who was notorious for his ambition and intrigues (See GSE, 1975, Vol. 8, pp. 204–205). Pëtr Andreevich Kleĭnmikhel' (1793–1869), a Russian statesman, General of Infantry (from 1841), Aide-de-Camp to A. A. Arakcheev from 1812, Chief of Staff of the military colonies from 1819, Nicholas I's Adjutant General from 1826, and a count from 1826, was implicated in large-scale corruption, which led to his dismissal in 1855 (See GSE, 1976, Vol. 12, p. 537). Vladimir Fedorovich Adlerberg (1790–1884) from 1817 was Aide-de-Camp to Grand Duke Nicholas (later Nicholas I). He was the Czar's closest friend during his entire reign. He became Adjutant General (1828), a count (1847), and Minister of the Imperial Court and Crown Lands (See GSE, 1973, Vol. 1, p. 10). — T.

46 K. A. Timiriazev: Awakening, p. 6. — A.

47 N. D. Zelinskiĭ: A. M. Butlerov i sovremennost' (A. M. Butlerov and the Present). Oktiabr' (October) *8*: 174 (1948). — A.

48 Borodin: Letters, III, pp. 127–128. — A.

[49] Istoriia Voenno-meditsinskoĭ (byvsheĭ Mediko-khirurgicheskoĭ) akademii za sto let (1798–1898). (History of the Military-Medical (former Medical-Surgical) Academy During the One Hundred Years (1798–1898)), St. Petersburg, 1898, p. 370. — A.

[50] See Appendix XI. — A.

[51] Letter of A. P. Borodin to A. M. Butlerov, March 23, 1880, Archives of the Academy of Sciences of the USSR, Stock 22, Inventory 2, No. 21, published here for the first time. — A.

Chapter 11

Prince Igor

At the end of the 1860's Borodin worked intensively in the chemical laboratory of the Medical-Surgical Academy. He gave only his free time to music. In 1869, inspired by the success of his First Symphony, he began to work on the Second (Bogatyr)[1] Symphony. At this same time he planned to write an opera.

V. V. Stassov, taking an active part in the search for topics for an opera, reported on this occasion:

> At the end of 1869 he [Borodin. — A.] planned his 2nd Symphony (B-minor) and played "material" from it to his friends. But, nevertheless, he proceeded to charge me with demands for a topic for an opera.
>
> He said that "he would like to create an opera more than a symphony." I made a new effort, and, under the influence of long conversations with him at our musical evening at L. I. Shestakova's[2] home on April 19, 1869, on that very night, I thought of a topic for an opera taken from "Slovo o polku Igoreve" (The Lay of Igor's Campaign). It seemed to me that here were all the objects necessary for Borodin's talent and artistic nature: vast epic motives, nationality, the greatest variety of characters, passion, and drama, each in its diverse manifestations. By the early morning of April 20th the very detailed adaptation was written, with excerpts from "Ipat'evskaia letopis'" (The Ipat'ev Chronicle) and from "The Lay of Igor's Campaign." I immediately sent my work and my explanation to Borodin. On the same day he answered me: "I do not know how to thank you for such an interest in my future opera . . . Your draft is so complete and detailed that everything comes out quite clearly . . . I like the topic immensely. Am I competent enough for it? I do not know. If you are afraid of wolves, don't go into the woods. I will attempt it." Thus on April 20, 1869 the fate of Borodin's opera was decided, which, as an epic opera, is similar to "Ruslan and Liudmila" by Glinka and is a superior epic opera of our century[3].

Becoming interested in the subject of the opera, Borodin asked Stassov to obtain literature elucidating the history of the campaign of the Prince of Novgorod-Seversk, Igor Sviatoslavich, against the Polovtsi.[4] Stassov, seriously occupied with the study of ancient Russian literary monuments, greatly helped Borodin. He obtained chronicles and literature from the Public Library on *The Lay of Igor's Campaign* in addition to the epic Russian songs, Zadonshchina and Mamaevo poboishche (The Battle of Mamai), and also Turkish and Finnish songs. Stassov also succeeded in obtaining for Borodin some native Polovetsian[5] songs, writen down by the Hungarian traveler, Paul Hunfalvi[6], partly in Asia and partly in one of the districts of Hungary where the descendants of the ancient Polovtsi lived. Studying the life of the ancient peoples from authentic documents of Russian antiquity, Borodin began to work enthusiastically on his planned opera.

However, as before, very little time was available to Borodin. Only during times of illness, which prevented his attendance at lectures and the laboratory, did he devote himself completely to music. Borodin wrote to Liubov' Ivanovna Karmalina:

> When I am so sick that I sit at home; when I am not able to do anything efficiently, my head is splitting, my eyes water, when I have to dig in my pocket for a handkerchief every two minutes – then I write music. As is the case now, I was also ill twice during the year, and both times the illness permitted the appearance of new bricks for the building of my future opera[7].

In 1870, having written Yaroslavna's[8] arioso[9] for *Igor* for the second scene of the first act (Ne malo vremeni proshlo s tekh por, Little time passes since that time), Konchakovna's[10] cavatina[11], and the procession of the Polovetsian princes, Borodin gradually lost interest in the opera. It seemed to him that he had "neither experience, nor ability, nor time" for carrying out this enormous work. Borodin wrote to his wife,

> Where have I really had anything to do with opera? The trouble and waste of time are enormous. The staging is questionable,[12] and even if they stage it, then where will I have time to mess with the whole string of petty troubles, annoyances with the management, with actors, with rehearsals, etc.? And meanwhile, the topic, no matter how great the music, will scarcely be able to please the public. There is little dramatic effect, and there is almost no stage movement. Finally, it is no trifle to prepare a libretto that satisfies the requirements both musically and dramatically. I do not have enough experience, ability, or time for this. The success of the opera is not guaranteed in any way. Erroneous mistreatment of the topic from a dramatic and staging aspect might come to light only afterwards, and to correct the affair would be just as difficult as with Ratcliff.[13] I arrived at all this after many attempts to compose some numbers out of the materials which are ready. Finally, an opera (that is not dramatic in a very strict sense) seems to me to be an unnatural thing. This was sharply pointed out to me after I heard "Proroka" (The Prophet)[14] on the Mariinsky stage.[15] I am by nature a lyricist and symphonist. I long for symphonic forms. For the present I will wait and write what will be written, not making up my mind to any large task[16].

Borodin turned all his attention to the Second Symphony. In 1871 he finished its first part. The composer's friends greeted the completion of the first part of the symphony with indescribable delight. Borodin reported to his wife,

> I was at Korsin'ka's [Rimsky-Korsakov's — T.]. I drank tea with him and sat with him for an hour or so, playing for him my new piece, with which he is delighted. In general, this piece is creating a noise in our ant hill. Cui came running early in the morning purposely to hear it[17].

How Balakirev reacted to the appearance of the new composition can be judged by one fragment from a letter of Borodin's to his wife:

> Miliĭ was too funny for words. I have already written to you that he has been sulking at me for some time and has been cold, angry, and from time to time nagging. I arrived at Liudma's [Shestakova's. — A.], and there was Miliĭ. You would hardly recognize him. He grew soft, looked at me with loving eyes and finally, not knowing how to express his love to me, gently took me by the nose between [his] two fingers and soundly kissed me on the cheek. I unintentionally burst out in laughter! You, of course, must have guessed the reason for such a change. Korsin'ka had told him that I am writing a symphonic piece and had played some of it for him. Funny Miliĭ![18].

Sometimes during the hours free from business and work in the laboratory, Borodin visited his musical colleagues. Most often it was a visit to Rimsky-Korsakov. Borodin wrote to his wife:

> Korsin'ka now lives alone ... He was indescribably happy for me. At once he ordered the samovar to be put on, and he himself began to make tea, and it was very funny: in a long jacket, awkward and beaming with joy, he swung his arms, shouted, brewed the tea, blew out the samovar, and poured. It is enough to make one split one's sides! I am terribly sad that you could not see him.
>
> We sat down to play; first two charming fugues of Bach, one of which I did not know at all (G Sharp Minor, in the 2nd book).[19] It was terribly good! It refreshed me very much after all the business, bustle, and fuss. Then he played your song[20]. Then I played for him my new symphonic thing which I am now concocting (which I played in Moscow). Korsets[21] raved and said that it is the most forceful and best of all my things. He shouted so and waved his arms, he stuck out his lower lip, blinked, and vamped now bass, now treble. Besides this we glanced over something else. I planned to stay only an hour and then visit my uncle. Instead, I listened for hours ... I count: one, two, three, four! ... And between the music we did not forget to drink tea, and the two of us remained sitting – [for] two samovars! For a long time I haven't had music like this my heart's content[22].

During the winter of 1871–1872 the director of the Imperial Theater, S. A. Gedeonov, proposed through Stassov to four of the composers of the Mighty Little Group, Borodin, Cui, Mussorgsky, and Rimsky-Korsakov, that they compose the music to the libretto, written by him, of an opera-ballet, *Mlada*. Planned as a grandiose and magnificent production, *Mlada* had to attract the public and assure large returns. The composers accepted the offer, and by the early spring of 1872 each of the four had finished his part. The subject of *Mlada* is concerned with the myth of the western Slavs, in particular, the Polabians.[23] Borodin worked on the fourth act of the opera, which attracted him by its epic grandeur. As was the case with the opera *Prince Igor*, Borodin studied a great number of works on the life, religion, and ceremonies of the Baltic Slavs.

Stassov wrote,

> In a short time Borodin created a series of scenes, which was amazing in respect to inspiration, profound historical color, and epic beauty ... All his

> comrades, who themselves created amazing scenes for "Mlada" at that time, were obliged to recognize the colossal, and in the present case, the overwhelming superiority of Borodin, and with a deep feeling of friendship and admiration they bowed before their gifted comrade[24].

However, the performance of *Mlada* did not take place. For a stage performance complex scenery and expensive costumes were required, for which it was necessary to spend a very great sum of money, and the management did not have it available.

Borodin later transferred the inspired and beautiful music of the scenes of the Apparition of the Phantoms of the ancient Slavic Princes from *Mlada* to the opera *Prince Igor* (the scene of the solar eclipse in the Prolog).

In the beginning of the winter of 1874 the young medical doctor V. A. Shanorov, a student of Borodin's, arrived at St. Petersburg from the Caucasus. Passionately loving Borodin's music, Shanorov began to protest ardently to his teacher that the opera must be finished. The beloved student's sincere enthusiasm appeared to Borodin as a stimulating push for resuming work on *Prince Igor*. In Stassov's

Fig. 12. Aleksandr Porfir'evich Borodin, portrait with autographed excerpt from his opera *Prince Igor*

words, "Borodin himself also again felt an appetite for his opera at this time."

During the summer of 1875 Borodin was with his wife in Moscow. Usually they moved to the country, most frequently in the picturesquely situated village of Davydovo[25] (now in Vladimir Province) along the Nizegorodskiĭ Road, where Borodin worked productively on the opera. In Moscow, in a burst of creative enthusiasm "with a lion's bite," according to Stassov's expression, he forged ahead on *Prince Igor*. He produced Khan Konchak's aria in the second act, the Polovetsian dances, [and] the first chorus in the first act (the scene with Vladimir Galitskiĭ).[26] Borodin wrote his wife upon his arrival back in St. Petersburg:

> The news of my musical activity in Moscow spread with lightning speed the very first day . . . To tell the truth, I did not even expect that my Moscow products would produce such a furor; Korsin'ka [Rimsky-Korsakov. — T.] is delighted, also Modest [Mussorgsky. — T.]. Liudmila Ivanovna [Shestakova. — T.] has invited Petrov to hear them. Enthusiasm for the first chorus especially amazed me . . . Without boasting I would say they found it especially effective, smart and deftly put over from a dramatic point of view. Konchak, it goes without saying, also makes the impression which I wanted . . . Of course, every one of them is only repeated over and over by the chorus so that I wrote the rest somewhat faster, on the spot. Cui came by purposely in order to hear it, but he did not find me at home . . . In general, this year no one rebukes me for inactivity on the part of music[27].

In 1876 Yaroslavna's recitative Kak unylo vse krugom (How Dreary is Everything Around Me) and some other numbers for the Fourth Act were added to the works which he composed in 1875.

At the end of this same year Borodin finished his *Bogatyr Symphony*, which was performed for the first time at the fifth concert of the Russian Musical Society on February 2, 1877 under Nápravník's[28] direction. "One of the mightiest and most capital compositions"[29] was coldly received by "society." The heroic symphony

Fig. 13. Borodin's autographed excerpt from the opera-ballet *Mlada*. Gift of A. P. Borodin to V. V. Stassov, March 5, 1872 (State Central Museum of Musical Culture, Moscow)

with a strongly pronounced, national, Russian character was not accepted because of the flippancy and bad taste of the aristocratic public of St. Petersburg. Musical "critics", who served the obscurantists from the Mikhaĭlovskiĭ Palace, did not rest until they slandered the remarkable creation of the great Russian composer.

While the malicious slander was raging, the progressive people of Russian society took in Borodin's *Bogatyr Symphony* with delight. The symphony was also an enormous success with famous European musicians. Their favorable reviews reached Borodin's ears and amazed him more than they made him happy. Borodin reported to his wife,

> Viardot[30] in Paris is delighted by my second symphony and has popularized it among musicians of the place. It is almost astonishing[31].

At the end of June 1877 Borodin went to Germany with his students, A. P. Dianin and M. Yu. Goldstein, in order to enroll them in Jena University to attain doctor of philosophy degrees.

In Jena Borodin learned that Franz Liszt[32] lived nearby in Weimar.[33] On July 3rd, on arriving in Weimar, Borodin met him. Borodin wrote to his wife,

> I did not even succeed in presenting my card when suddenly before my nose, as though out of the earth, a tall figure appeared in the entrance in a long black frock coat, with a long nose and long gray hair. "You composed an excellent symphony," the figure shouted with a loud voice, and a long arm was extended to me. "Welcome. I am delighted . . . The first movement is excellent, your andante is a masterpiece, the scherzo is exquisite, and then this passage is so ingenious . . ." And on and on. His strong hand firmly pressed my hand, and he made me sit down on the sofa. I could only nod approval and thank him. The majestic figure of the old man, with an energetic, beautiful face, animated, moved in front of me and talked without stopping, plying me with questions concerning my personal and musical affairs in Russia, which evidently were rather well known to him. The conversation proceeded now in French, now in German, skipping continually from one to the other. When I told Liszt that I am, strictly speaking, "an amateur musician" (really a "Sunday musician"), he even made the pun: "but you know Sunday is always a holiday, and you have an absolute right to 'feiern,' that is to celebrate." He was terribly pleased with my piano arrangements and said that they reveal in me a musician who is "experienced and extremely talented, possessing a contemporary piano technique"[34].

Borodin asked Liszt what defects were present in the symphony and reported to him that several musical critics had berated his symphony. Liszt answered:

> "Please do not listen to those who would hold you back from your course; believe in yourself that you are on the true path. You have so much artistic flair that you need not be afraid to be original, and remember that exactly such advice was given in their time to both Beethoven and Mozart and others, and they never would have become great masters if they had taken it into their heads to follow it." When I thanked him for the compliment, he interrupted me with

annoyance: "But I am not complimenting you; I am so old that it is not appropriate for me to say to someone something other than what I think. They do not love me here for this, but I am not able to say that here they write good things when I find them bad, lacking in talent, and lifeless. They say," Liszt continued, "that there is nothing new under the moon, but you see this is absolutely new. You would not find this in any other composer." The great "Maestro" reiterated with regard to various details. "A German was with me yesterday," Liszt said. "He brought me his *Third* Symphony. I said to him: We Germans are still a long way from this, and I showed him your Second Symphony. You know Germany? Here they write much. I am lost in the sea of music which threatens to submerge me, but good God, how flat it all is. Not one fresh idea! With you a living stream flows; sooner or later (no doubt, later) it will also force its way into Germany"[35].

Afterwards, developing this same idea, Liszt said to Borodin:

"Look! In our country they write in this manner! Look! Well, what sort of thing is it? . . . Well, is it not really the most trivial Mendelssohnism! And here in Germany they constantly treat us to such music! Here, wait a little, you will hear all this yet today! No, you Russians are indispensable to us. Without you I am powerless — vous autres Russes [you Russians]," Liszt said, smiling. "You have a lively, vital stream; you have a future, but around here there is dead flesh for the most part"[36].

While the "boys" energetically worked on their dissertations, Borodin attended Liszt's concerts and was often together with the famous musician. Sincerely sympathizing with the new Russian musical school, Liszt highly appraised the symphonies of the talented Russian composer. He did much to popularize abroad the music of Borodin and his comrades from the Mighty Little Group. Works of Glinka and other Russian composers were also performed at many concerts upon Liszt's advice.

In 1880 (September 3) Liszt wrote to Borodin from Rome:

I am very late with my statement concerning what you must know better than I; it is this: the instrumentation of your highly remarkable symphony (B Minor) was made by the hand of a master and corresponds admirably to the composition. For me it was a real delight to hear it at rehearsals and at the concert of the "Musical Congress" at Baden-Baden. The best authorities and the multitudes of the public acclaimed you[37].

Borodin left for Russia on July 21st. He was at ease concerning his students inasmuch as their dissertations fully merited doctor's degrees. Borodin wrote to his wife:

Geuther showed me a lot of dissertations which were done at the Jena laboratory. I can say with pleasure that they all are not fit to hold a candle to Pavlich's [Dianin's. — A.] dissertation, and in most cases they are weaker than Goldstein's dissertation[38].

On the way to Marburg Borodin visited the Third Meeting of the Marburg Society of Naturalists where he heard Haeckel's[39] report on polyps and jellyfish.

On the way Borodin also visited Heidelberg where he visited the places dear to him in his reminiscences. In this small town he had begun his scientific life 17 years earlier; his first time of happiness had been there.

Having returned to his native land, Borodin continued to work actively in the chemical laboratory of the academy and with the women's courses.

In 1879 he finished the First Quartet in A Major.[40] During the summer of 1879, resting in Davydovo, Borodin worked a great deal on the libretto of the opera. After careful study of the Ipat'ev Chronicle he significantly altered Konchak's role and wrote small parts for separate scenes of the opera.

Shortly before this, Borodin composed some humorous songs for piano duet on the theme of a children's polka: Polka, Requiem, Funeral March, and Mazurka, published in a series of *Parafrazy* (Paraphrases)[41], written by him together with Cui, Anatoliĭ Konstantinovich Liadov, and Rimsky-Korsakov. On July 15, 1879 Liszt sent the following letter from Weimar to the authors of *Paraphrases*:

> Esteemed gentlemen, you created a composition with a serious significance in the form of a joke. Your "Paraphrases" delights me: there is nothing wittier than your 24 variations and 14 small songs . . . Here, finally, is a first-class, concise guide to the science of musical harmony, counterpoint, rhythms, [and] fugue-like style. I will willingly propose to professors of composition of all the conservatories of Europe and America to take your "Paraphrases" for practical guidance in their teaching . . . I thank you, gentlemen, for this entertainment, and when one of your writes some new composition, I beg him to report this composition to me. The most lively, lofty, and sympathetic respect on my part has for many years already belonged to you. Franz Liszt[42].

Although domestic anxieties also distracted Borodin from his activity as a composer, nonetheless, he continued to work on *Igor*. Borodin wrote to Stassov,

> For the 2nd scene I wrote a duet for Vladimir Galitskiĭ and Yaroslavna, a women's chorus, and a small scene with Yaroslavna. Vladimir Galitskiĭ is boldly taking shape in all this. Now he appears as a small but rather bold role. In all cynicism I made him a prince and not too coarse, for that would be a second copy of Skula.[43] He is simply a nasty gamin, cynical but not devoid of some elegance, and not quite a cruel tyrant. Now I am working on the finale of the 1st Act. There I made a digression from the original plan of the libretto, for which you, perhaps, will take a stand to reproach me. Namely, I did not make the merchants themselves report the news about Igot, the utter defeat at Kalka,[44] and so forth, but boyars[45] and the retinue of the princess who had found out the terrible news earlier from the merchants. They arrive to prepare the princess in order not to stun her by the news, and when she finally finds out about the affair, she faints away, regains consciousness, and asks: who transmitted the terrible news? They tell her visitors on tour, and she orders to meet them to question them in detail about everything. This was done in order to avoid the tale of the merchants. The motives for this are the following. If the tale were made complete,

> poetic, and picturesque, as in the original, then it would be long and boring, and in the end one would not hear all of the words. It is not possible to give grandiose music which would be proper for the circumstance to mediocre singers (for first class ones are used for other roles) so all of this would turn out to be worse than with the chorus. From the libretto side there was always one effect, that the merchants narrate, "interrupting one another", but this effect is purely superficial[46].

This fragment from the letter to Stassov shows with what thoroughness Borodin thought out and trimmed each detail of his opera. He nurtured all his other compositions for a long time, and only when he was convinced of the full artistic value of his creation did he show it to his comrades. Not being completely content with the scenario of *Prince Igor*, which was suggested by Stassov, Borodin did much work on the text. This resulted, as Stassov observes, in the fact that the opera became all the more beautiful.

During these years Borodin occupied himself with work on the opera almost exclusively in the summer in the country, where he brought his piano. During the winter in St. Petersburg one could not talk about musical creation. Borodin reported to his wife,

> Now at home I am completely absorbed in the pool of academic life, examinations, reexaminations, lectures, the laboratory, the Conference, committees, reports, messages, all this immediately envelops me . . . But I, like a regimental horse hearing a trumpet, prick up my ears and rush with fresh strength into the service of academic activity[47].

Borodin clearly realized that he could not find the necessary time for the creation of Prince Igor. On one hand, he wanted to finish his opera, but on the other, he was afraid to take too great an interest in it for fear that it would have a harmful effect on his scientific activity.

In the end such ambivalence had the result that the opera *Prince Igor* was not finished by the composer.[48] However, by 1887 Borodin had already composed all the basic musical material of the opera. His untimely death wrecked his plans. After his death Rimsky-Korsakov and Glazunov put the numerous fragments of writing into order and united them into a whole, from which time Borodin's opera has entered the treasury of Russian operatic classics.

As the greatest Soviet musical expert, Academician Boris Vladimirovich Asaf'ev, justly noted,

> A national-state conception is contained in this opera . . . The thesis of a sovereign construction as defense against the elements of the steppes with its nomads and against feudal-princely destruction and anarchy is carried out by a powerful, fiery, one could say Surikovian,[49] hand throughout the entire opera, and namely this realm of substance of "Igor" substantiates the monumentality of the music . . . Borodin boldly and lushly compares two worlds of state systems: that of the slaves, the nomads, the violent – the Polovtsi, and that of the defensive, laboring, self-protecting regime of the Great Russian toiler – the peasantry and the armed troops[50].

The ideological-artistic significance of *Prince Igor* depends first of all on the patriotic direction of the opera, which arises from the national springs of the Russian historical epos.

> The composer understood that in the opera, first of all, the themes of the patriotism of the Russian people must be heard, the theme of the unification of the Russian land, to which the call resounded both in the chronicles and in the heroic poetry of the Russian middle ages. From here stems the slow, concentrated, zealous tempo of the unwinding of events in "Prince Igor;" from here comes the majestic simplicity of the musical-dramatic narrative, sincerity, and harmony of thought; from here comes the reception of the introductory hymn which, in essence, is the prolog of the opera[51].

Stassov wrote,

> The national element constitutes the most forceful note in Borodin's creative work. He did not yield to either Glinka or to his comrades of the new Russian musical school in the truthfulness and depth of the Russian character of his superior creations[52].

Prince Igor was the first Russian heroic-epic opera. Developing the patriotic traditions of Glinka's *Susanin*,[53] Borodin maintained in his opera the idea of love for one's native land, which took shape during the course of many centuries, and of readiness to give one's whole life to it, which was the case in the wars of the princely detachments at the Kaiala River, and as Igor himself was prepared to do. It is quite natural that this idea is embodied in the opera by broad, expressive means, similar to Russian folk music, free and powerful.

Borodin's splendid lyrical talent is clearly expressed in *Prince Igor*. That heartfelt sincerity, which is especially characteristic of creative Russian folk songs, becomes apparent in the aria of Igor,[54] who in Polovetsian captivity remembers his "dear Yaroslavna," and in the moving *Plach Yaroslavny* (Yaroslavna's Lament).[55] It is impossible not to recall Russian folk lamentations when listening to Yaroslavna's Lament. And along with the revelation of the inexhaustible wealth of Russian song, beginning with its epic power and ending with the fine lyrical women's grief, in *Prince Igor* Borodin creates, with musical riches, another world, which appears before us in the scenes of the Polovtsian camp,[56] in the famous *Polovetskie Pliaski* (Polovtsian Dances), in the procession of the Polovtsian princes, in the lyrical effusion of Khan Konchak's daughter, who fell in love with the captive son[57] of the Russian prince, and in other images of the Orient[58]. Images of the Orient also spring up in the song *Arabskaia melodiia* (Arabian Melody)[59] (to his own text) and in the symphonic poem *V Sredneĭ Azii* (In Central Asia).[60] Here Borodin, preserving all the originality of his creative aspect, resembles Glinka and his comrades of the Balakirev Circle.

The epic forms of *Prince Igor*, much akin to Borodin's Bogatyr Symphony, according to the words of the composer himself, were born from the scenes of the Russian epos. If we want to name the most brilliant, the most characteristic trait of Borodin's creative works, then, naturally the first choice falls upon the Herculean, epic power of his music, which is shown not only in *Prince Igor* and in the

Bogatyr Symphony but also in the First Symphony, in the ballad *More* (The Sea), and in the most brilliant *Pesnia temnogo lesa* (Song of the Dark Forest). This power not only reflects the originality of Borodin's tremendous talent but also embodies one of the most basic traits of the national aspect of the great, powerbul, and invincible Russian people.

Footnotes to Chapter 11

[1] Bogatyrs were legendary Russian heroes, of whom Il'ia Mourometz, the subject of Reinhold Glière's Symphony No. 3, in B Minor, Op. 42, is among the most famous. — T.

[2] Liudmila Ivanovna Shestakova (1816–1906), a sister of M. I. Glinka's, was a well-known musical-social figure and was close to the Balakirev Circle. — A.

[3] V. V. Stassov: Biography, p. 197. — A.

[4] The Polovtsi or Polovetzki (called Kipchaks by the Turks and Kumans by the Byzantines) were a confederation of nomadic tribes who, by the middle of the eleventh century, occupied a vast, sprawling territory in the Euro-Asian steppe, stretching from north of the Aral Sea westward to the region north of the Black Sea. In 1237 Mongol invaders killed Bachman, chief of the eastern Polovtsi tribes, and the confederation was destroyed. — T.

[5] In English, the adjectival form is rendered variously as Polovetsian, Polovtsian, Polovetzian, etc. — T.

[6] Paul Hunfalvi (1810–1891) was a Hungarian linguist and ethnographer. — A.

[7] V. V. Stassov: Letters, p. 4. — A.

[8] Prince Igor's wife by his second marriage. — T.

[9] A short air in an opera or oratorio. — T.

[10] Daughter of Konchak, Khan of the Polovtsi. — T.

[11] An operatic solo aria or song-like air. — T.

[12] The action of the Prolog and of the First and Fourth Acts takes place in Putivl, a city about 160 miles (ca. 257 km) northeast of Kiev, and that of the Second and Third Acts in the Polovtsian camp in the year 1185. — T.

[13] See footnote 20, Chap. 7. — T.

[14] Probably, *Le Prophète*, an opera by Giacomo Meyerbeer, dealing with the sixteenth-century uprising of the Anabaptists in the Netherlands. — T.

[15] The Mariinsky Theater has been renamed the Kirov State Academic Theater of Opera and Ballet. — T.

[16] Borodin: Letters, I, pp. 200–201. — A.

[17] Ibid., p. 227. — A.

[18] Ibid., p. 221–222. — A.

[19] Das wohltemperierte Klavier (The Well-Tempered Clavier) consists of two books (Book 1—1722; Book 2—1742), each containing 24 preludes and fugues in all keys. The fugue in question is No. 18 in the collection. — T.

[20] Rimsky-Korsakov's song V tsarstvo rozy i vina pridi (Come into the Kingdom of Roses and Wine), dedicated to Borodin's wife, Ekaterina Sergeevna Borodina. — A.

[21] Rimsky-Korsakov. — T.

[22] Borodin: Letters, I, pp. 219–220. — A.

[23] This tribe lived in the Elbe (Laba) River valley. The extinct Polabian language was spoken by the Slavic population of this region until the seventeenth or eighteenth century. — T.

[24] V. V. Stassov: Biography, p. 159. — A.

[25] Davydovo is located about 15 miles (ca. 24 km) east of Vladimir, which in turn is situated about 115 miles (ca. 185 km) east of Moscow. — T.

[26] Yaroslavna's brother and Prince Igor's brother-in-law. — T.

[27] Borodin: Letters, II, pp. 98–99. — A.

[28] Eduard Frantsevich Nápravník (1839–1916), a famous conductor and composer, was conductor of the Mariinsky Theater in St. Petersburg from 1867. From the autumn of 1869 to 1881 he conducted the concerts of the Russian Musical Society. The most famous of Nápravník's pro-

ductions are the operas Dubrovskiĭ and Nizhnegorodtsy (The People of Nizhnegorod). — A. [See Jennifer Spencer: Eduard (Francevič) Nápravník. In: Grove, Vol. 13, pp. 35–36. — T.]

29 V. V. Stassov: Sobranie sochineniĭ (Collected Works) [hereafter abbreviated as Collected Works], Vol. 3, St. Petersburg, p. 314. — A.

30 Pauline Viardot (1821–1910) was a famous singer and a friend of I[van] S[ergeevich] Turgenev's in Paris. — A. [See April Fitzlyon: (Michelle Ferdinande) Pauline Viardot. In: Grove, Vol. 19, pp. 694–695. — T.]

31 Borodin: Letters, II, p. 191. — A.

32 Franz Liszt (1811–1886), a well-known Hungarian composer and virtuoso pianist, was one of the founders of Western European program music. Liszt was a friend and popularizer of Russian music (see the work of V. A. Kiselev: Frants List i ego otnoshenie k russkomu iskusstvu (Franz Liszt and His Attitude toward Russian Art), Moscow, 1929). Borodin met with Liszt three times — in 1877 and 1885 in Weimar and in 1881 in Magdeburg. — A. [See Humphrey Searle: Franz [Ferenc] Liszt. In: Grove, Vol. 11, pp. 27–84. — T.]

33 Weimar is about 12 miles (ca. 19 km) northwest of Jena. For additional information on Borodin's relations with Liszt see Louise Cruppi: Borodin and Liszt. Living Age, *312*, 600–605 (March 11, 1922) and David Lloyd-Jones: Borodin on Liszt. Music and Letters; *42* (2), 117–126 (1961). Also see the book listed as No. 23 on Page 133. — T.

34 V. V. Stassov: Letters, pp. 11–12. — A.

35 Ibid. — A.

36 Ibid., p. 164. — A.

37 V. V. Stassov: Biography, p. 164. — A.

38 Borodin: Letters, II, p. 157. — A.

39 Ernst Haeckel (1834–1919) was a famous German naturalist and professor at Jena University. — A. [See Georg Uschmann: Ernst Heinrich Philipp August Haeckel. In: DSB, 1972, Vol. 6, pp. 6–11. — T.]

40 Dedicated to Rimsky-Korsakov's wife, Nadezhda Nikolaevna Rimskaia-Korsakova (née Purgold). — T.

41 Borodin conceived the idea of writing *Paraphrases* during the 1870's (most probably in 1874), when a young girl, Gania Litvinenko (see footnote 9, Chap. 12), who later became his adopted daughter, asked him to play piano duets with her. When he said that he did not think that she could play, she exclaimed, "Oh, yes, I can; Look, I can play this!" and played "Chopsticks" with two fingers (Serge Dianin: Borodin, Oxford University Press, London, New York, 1963, p. 247). — T.

42 V. V. Stassov: Collected Works, Vol. 3, p. 501. — A.

43 Skula (bass) and Eroshka (tenor) are villainous gudok players in the opera. The gudok was an ancient three-stringed fiddle of Russian and Belorussian (White Russian) origin. It had a pear-shaped body about 32 inches (ca. 80 cm) long and was played while held on the knee. — T.

44 Some investigations of *The Lay of Igor's Campaign* indicate that for the Kaiala River one should understand the river Kalka. — A.

45 Members of the old nobility of Russia before Peter the Great made rank dependent upon service to the state. — T.

46 V. V. Stassov: Letters, p. 29. — A.

47 Borodin: Letters, II, p. 80. — A.

48 Jean-Albert Gautier contended that Kekulé's discoveries prevented Borodin from completing *Prince Igor* (Comment les découvertes du chimiste Kekulé empêchèrent Borodine de terminer 'Le Prince Igor.' Revue d'Histoire de la Pharmacie. *20*: No. 204, 5–10 (March, 1970)). For additional information about the opera see Gerald Abraham: Studies in Russian Music, William Reeves, London, [1936], Chap. 7, "Prince Igor" (pp. 119–141); Ward Botsford: Angel's 'Prince Igor', American Record Guide. *34*: 358–361 (January, 1968); and V. A. Kiselov: Stsenicheskaia istoriia pervoĭ postanovki Kniazia Igoria (The History of the First Performance of Prince Igor). In: Mikhail Pavlovich Alekseev et al. (eds.): Muzykal'noe nasledstvo (Musical Heritage), Moscow, 1970, Vol. 3, pp. 284–352. — T.

49 Vasiliĭ Ivanovich Surikov (1848–1916) was a Russian painter of historical scenes. See GSE, 1980, Vol. 25, p. 245. — T.

50 Igor Glebov (B. V. Asaf'ev): 'Kniaz' Igor' — opera Borodina (Prince Igor — Borodin's Opera). In: Sovetskaia muzyka (Soviet Music), second collection of articles, Muzgiz, [Moscow], 1944, p. 17. — A.

[51] Natsional'nye istoki tvorchestva Borodina (Nationalistic Sources of Borodin's Work). In: Nauchno-tvorcheskaia rabota VTO (Scientific-creative Work VTO), Moscow, 1948, Vol. 3, p. 34–35. — A.
[52] V. V. Stassov: Borodin, p. 29. — A.
[53] See footnote 1, Chap. 7. — T.
[54] Act II, No. 13. — T.
[55] This aria opens Act IV (No. 25). — T.
[56] Acts II and III. — T.
[57] Vladimir Igor'evich. — T.
[58] For Borodin's paternal oriental heritage see footnote 6, Chap. 1. — T.
[59] For further details see Gerald Abraham: Arab Melodies in Rimsky-Korsakov and Borodin. Music and Letters *56* (3–4): 313–318 (1975). — T.
[60] This symphonic poem first established Borodin's reputation outside of Russia. — T.

Chapter 12

Last Years

The 1880's were years of furious reaction in the history of Russia, which were shown first of all in the brutal suppression of the revolutionary movement. Everything that was advanced and progressive in Russian science and art was persecuted and harassed.

A letter from Professor V. Ya. Danilevskiĭ[1] of Kharkov University to A. M Butlerov clearly characterizes the oppressive atmosphere of this time:

> Such a severe, leaden storm cloud, such an oppressive atmosphere envelops all our professors at Kharkov University and poisons all our academic-educational life. Yes, and not only us, one can also hear complaints from other universities. But what is going on here with us, thanks to the presence of two or three members of the Council who are indeed morally leprous, it seems to me, surpasses all probability . . . And this is done in the name of, under the aegis of a new banner! All the best and not best people feel and understand exactly the same. Neither fear nor fright of "Quos ego" [I will give you "what-for." — T.] of the power, nor fear for personal safety, nor fear of spies and denunciations (we are already disciplined to this) call forth our lamentations as an expression of our personal sensations . . . In society outside the university they talk just the same.
>
> For example, at the last Council meeting anyone from the outside would certainly regard it as a meeting of some violent, groveling rabble, requring moral and mental treatment.

Speaking of the raving reactionary scholars, Danilevskiĭ further wrote that

> formerly almost none of them were heard; now by brutal, savage instinct they feel *solid* ground underfoot, and they tear and interfere with everything that is mutual, autonomous, or hated by them; everything that formerly restrained their violent instincts . . . These spokesmen "of the triumphant pig," according to Shchedrin,[2] belong entirely to that newest structure which was so correctly presented in the newspaper Russkie vedomosti (Russian News) on January 26, 1885 (No. 25).
>
> We now have a grave atmosphere; the only refuge is this laboratory and its office. But even here we do not always find peace, even here there are rumors, fears, and interrogations[3].

The reactionary scholars in the universities and the academy raised their voices, which until now had been muffled by the social development of the 1860's. Zealously serving the governments of the landlord-noblemen, these scholars promoted the suppression of everything lively and new in Russian science. With servile delight they received the circular about "kukharkinye deti" (the maid's children) published by the government during this period. According to the circular, one ought not to

> withdraw the children of coachmen, footmen, cooks, laundresses, small shopkeepers, and "similar people" from the environment to which they belong. This, as many years of experience show, brings them to disregard their parents, to be dissatisfied with their mode of life, to be bitter against the inequality of the property status which exists and is inevitable according to the very nature of things[4].

In 1884 new university regulations were established by the Minister of Education, Ivan Davidovich Delianov, according to which self-government in universities was abolished, professors and rectors of universities were appointed by the government, etc. In the beginning of the 1880's the Czarist government discontinued a group of progressive journals, on the pages of which the reactionary politics of the autocracy were caustically castigated. Suppressing the social movement, the Czarist government discovered evidence of "sedition" in the very existence of the women's courses. In 1882 admittance to the women's medical courses was discontinued, and in 1885 the courses were terminated.

Borodin, who during the course of many years had created his beloved child, the women's higher courses, with enormous difficulty, was shocked by its death.[5] Until the last day of his life he did not lose the hope that the courses would return to life again. "It is not possible that the work of women's medical education in Russia should perish so," he said. We have already showed that in the period of the organization of the courses the better part of Russian society responded enthusiastically to the calls of Borodin and his colleagues. The leading professors of Kiev and other Russian universities helped the women's courses materially; they read lectures on their behalf, carried out collections for the courses, etc.

In one of the letters to an unknown person who had rendered support to the courses (we were not able to ascertain the name of the addressee) Borodin wrote:

> Allow me to convey to you the most hearty thanks for your sympathy and active help to our women's medical courses. This sympathy is especially dear now, when the courses must experience perhaps the most difficult time of their existence. Ten years have passed since their establishment; nevertheless, they already have an excellent past which was granted to them by the respect and sympathy of the better part of Russian society. There is honor and glory for that part of society which without any official call hastens to the aid of a young establishment and does its bit in order to rescue the perishing courses from misfortune. And it must be confessed that the future of the courses now depends entirely upon the material support of society[6].

However, neither the remarkable donations on behalf of the women's courses received from various corners of Russia, nor the active part in their work of the country's most prominent scholars, nor the efforts of Borodin and other figures of women's education were able to save the courses.

The Czarist government opposed by every means the diffusion into Russian society of any materialistic ideas which were advocated by the foremost Russian scholar-naturalists. Not confining itself to the discontinuation of journals, to lashing censorship, and to depriving the best scholars of university positions, the Czarist government, during the 1880's, was not squeamish about other "methods." It heavily curtailed allowances for laboratory work, purchase of new laboratory equipment, reagents, dishes, materials, etc. In one letter Borodin reported:

> With us in the Academy the curtailment of expenses has also reached the ridiculous. They took away the gas. They are restricting the number of attendants to an impossible degree[7].

The suffocating atmosphere of reaction and the death of the courses undoubtedly reflected on Borodin's creative activity as a chemist. During the last years of his life (1880–1887) he published vitually no chemical research. Of course, his extremely busy work on countless committees of which he was, as a rule, an indispensable member, and his active part in charitable societies, etc. also adversely affected him. Borodin wrote to his wife,

> No, it is difficult to be simultaneously both Glinka and Semen Petrovich,[8] both a scholar and an agent, both an artist and an official, both a philanthropist and a father of someone else's children[9], both a doctor and a patient . . . It looks as if I will end up only as the latter. I am not only leaving from here for the country but it seems for the devil . . . You cannot imagine how angry and irritable I am now[10].

Of course, under such conditions, when "one affair urges on another," it was difficult to find time for systematic and extended scientific research. Borodin wrote to Karmalina:

> We sinful ones spin around, as before, in a whirlpool of everyday, official, academic, scholarly, and artistic bustle. We hurry everywhere and do not have time anywhere; time flies like a locomotive at full speed[11].

However, even though he himself was not involved in scientific research during this period, Borodin did give much attention to the guidance of the scientific work of his students in the chemical laboratory of the Medical-Surgical Academy.

During 1879–1880 Borodin gave several reports on the chemical work accomplished in his laboratory at meetings of the Russian Physico-Chemical Society.[12] On January 11, 1879 he reported on P. G. Golubev's behalf that the product of the reduction of nitrobenzoin, obtained by him in 1874, proved to be aminodesoxybenzoin.[13] This product was obtained by the reduction of nitrodesoxybenzoin by tin in hydrochloric acid.[14] On February 1, 1879 Borodin reported on melissic acid[15] on behalf of M. Shalfeev.[16]

On November 6, 1880 Borodin gave three reports: on N. N. Bronnikov's work: Kalorimetricheskiĭ sposob opredeleniia ammiaka i azotnoĭ kisloty (A Calorimetric Method for the Determination of Ammonia and Nitric Acid), on P. Golubev's work:

O dinitrodesoksibenzoine i ob okislenii ikh khromovoĭ kislotoĭ (On Dinitrodesoxybenzoin and its Oxidation by Chromic Acid),[17] and on behalf of Master of Pharmacy [A.] Drygin (from Kytaisi[18]): O novoi dvoinoĭ soli khloristovodorodnogo khinina i khloristovodorodnoĭ mocheviny (On the New Double Salt of Quinine Hydrochloride and Urea Hydrochloride).[19]

In 1881 Borodin completed his 25th year at the Medical-Surgical Academy. This gave him the right to a pension. During this period he wrote a report addressed to the head of the Academy.[20] A student of Borodin's, Mikhail Yul'evich Goldstein, wrote in 1885:

> He [Borodin. — A.] has now completed his scientific activity and has the right to do so ... He has devoted himself completely to general musical activity and, unfortunately, still also to public activity which encroaches on his time.[21]

In the years already preceding this, both of Borodin's symphonies and separate numbers from the opera *Prince Igor* had been performed with great success in concerts in St. Petersburg, Moscow, and Kiev. During the winter of 1880 Borodin wrote his wife from Moscow:

> The symphony went well, and, to my surprise, after each part they clapped and called so that I had to go out on the stage and give two bows. The orchestra also applauded boisterously. The professors of the conservatory congratulated [me] and paid me a lot of compliments. When I went home after the concert a group of ladies was talking behind me: "It is he, it is he! Borodin! Follow him, see where he is going!" ... In short, I have now been more or less presented to Moscow's musical society[22].

However, having finished Igor's remarkable aria [Act II, No. 13. — T] in 1881, Borodin worked very slowly on the opera in the following years. An acquaintance of Borodin's, M. M. Kurbanov testified:

> None of the admonishments of Borodin's closest friends Rimsky-Korsakov, Stassov, Cui, and others, nothing was able to push for the successful completion of the highly original opera, which showed the composer's great talent. Everything hampered Borodin, and one disorderly, restless day after another took its course, not giving him one free minute for composing[23].

Rimsky-Korsakov also wrote about this:

> His swarming engagements in connection with his professorship and medical courses for women, were always in the way. His home life I have already described. Owing to his infinite kindliness and his entire lack of self-love, these surroundings made it extremely inconvenient for him to work at composition. One might come again and again and keep demanding how much he had written. Net resilt – a page or two of score, or else – nothing at all. To the query: "Aleksandr Porfir'evich, have you done the writing?" he would reply: "I have." And then it would turn out that the writing he had done was on a batch of

> letters! "Aleksandr Porfir'evich, have – you – finally – transposed such and such a number of the opera score?" – "Yes, I have." – he replies earnestly. "Well, thank the Lord! at last!" – "I transposed it from the piano to the table." – he would continue with the same earnestness and composure![24]

In a letter to Karmalina, Borodin wrote about this same period:

> You ask about "Igor". When I talk about it, it becomes ridiculous to me. I somewhat resemble the Finn in Ruslan.[25] As he, in dreams of his love for Naina, not noticing that time is passing, solved the problem when both he and Naina have grown old and gray, so I too strive to realize a cherished dream – to write an epic Russian opera. But time flies with the speed of an express train: days, weeks, months, winters pass under conditions not permitting one to think about serious musical work. It isn't the fact that I can't find an hour or two of idle time in the day! No, it's the fact that I do not have the moral leisure time. There is no possibility of brushing aside the flock of daily anxieties and thoughts, which do not have anything in common with art, which constantly thrive and swarm before you. There is no time to change one's mind, to attune oneself in musical harmony, without which creativity in large things such as opera is impossible. I have had at my disposal such a frame of mind only part of the summer. In the winter I can write music only when I am so sick that I do not read lectures, do not go to the laboratory, but, for all that, am able to do something. On this basis my musical comrades, in spite of conventional custom, always wish Everyone spoke highly flatteringly about "Antar" and praised it. It was a positive able to work in the laboratory. I sat at home and wrote the chorus of celebration for the last act of "Igor." Just as during the time of a mild illness I wrote "Yaroslavna's Lament,"[26] etc.[27].

During the spring of 1881 Borodin went to Magdeburg where at the Eighteenth Congress of Musicians of the Allgemeine Deutsche Musikverein (All-German Music Society) he met Liszt for the second time. Here under Nikisch's[28] direction, Rimsky-Korsakov's *Antar*[29] was performed with great success. Borodin wrote to his wife,

> Everyone spoke highly flatteringly about "Antar" and praised it. It was a positive success . . . In general, the stock of our group rates very highly here, higher than I thought. My symphony gave me such an honorable reputation in musical circles in Germany that I cannot give my name without greetings and the warmest praises following[30].

Borodin remained in Magdeburg until the end of June, then he went to Weimar, where, as he writes, apart from the "gray Venus"[31], the latest presentation of Goethe's *Faust* fascinated him. In Weimar Borodin showed Liszt his symphonic poem V Sredneĭ Azii (In Central Asia), with which Liszt, an ardent admirer of Russian music, was delighted. Borodin wrote to his wife: "I dedicated 'Central Asia' to Liszt. He kissed me affectionately and thanked me profusely"[32].

In July, 1881 Borodin returned to Russia. He wrote to Dianin after returning from abroad:

> I carried out and concluded little with respect to chemistry in the sense that with our lack of money at the Academy one cannot think about the introduction of apparatus, etc. that I could see[33].

In the following years (1881–1883) Borodin composed some songs, among which Dlia beregov otchizny dal'noĭ (For the Shores of a Distant Homeland)[34] with words by Pushkin was singled out as one of the best Russian songs.

During 1881–1882 Borodin wrote his second quartet,[35] and in 1885 he finished the Malen'kaia siuita (Petite Suite)[36] for piano. Kurbanov wrote:

> With regard to this suite I happen to know the following. The impossibly dirty sheets of various sizes of note paper with many corrections, on which the suite was composed, were left forgotten by him [Borodin. — A.] in the Argenteau castle[37] upon his departure to Russia. In the following year, 1886, when he again found himself in Belgium (where his creations performed there at that time were met with extraordinary ovations), he saw an unusually luxurious bound book cover on the table in the drawing room of the Argenteau castle. Opening it, to his astonishment, he saw his forgotten dirty pages of the piano suite on the pages of a rough notebook[38].

This small touch shows with what great respect Borodin's productions were received abroad.

In 1885 Borodin also wrote a piano scherzo.[39] In the same year he received the first printed excerpts from his opera *Prince Igor*, which were published in Paris.

In August, 1885 Borodin went abroad to participate in Russian concerts in Liège, Belgium. In Weimar he met with Liszt for the third time. This meeting was the most heartfelt. Four pieces, written by Borodin at the request of the Countess de Mercy-Argenteau, were performed by Liszt and received his high appraisal. After staying in Weimar for only 6 days, Borodin went to Antwerp and then to Paris and Liège, where he was met with ardent ovations. Belgium and France hailed the great Russian composer and made feasts in his honor.

In Antwerp Borodin received invitations to conduct two concerts at the Antwerp Exhibition. Borodin wrote his wife,

> I resolutely declined to conduct concerts, alluding to the lack of time for preparation of the concerts, to the absence of the habit of conducting large concerts, and finally, to the fact that in general I am not a conductor by profession. I am able to decline the conductorship, but I am not able to decline my presence at the concerts where my creations are performed . . . It would have been simply foolish and harmful for me as a musician and for all Russian music to flee from there . . . It was reported to me that a great number of people were ardently interested in my music and came *purposely* in order to become acquainted with my music and with me personally[40].

Indeed, lovers of music, composers, and professors of the conservatory from Liège, Brussels, Ghent, Holland, and France arrived at the concert. Borodin wrote to his wife,

> I note a characteristic fact. *The German party* in Belgium (constituting there the same evil as we have) exhausted every possible and impossible means in order to prevent the performance of my second symphony in Antwerp! Is it not true that this is extremely characteristic? Nevertheless, nothing further stood in the way: "Central Asia" was performed on September 9th, the symphony was performed on September 16th, and the first symphony was performed on September 19th. Besides this "More" [The Sea], "Morskaia tsarevna" [The Sea Princess],[41] and "Spiashchaia kniazhna" [The Sleeping Princess][42] were performed[43].

The triumph of Russian music, in spite of the efforts of all those who disparaged its grandeur, was clearly expressed in the passionate ovations of the numerous guests of the concerts in honor of its representative, Aleksandr Porfir'evich Borodin.

At the end of September, 1885 Borodin returned to his native land. At the end of 1886 he began the third symphony[44] for full orchestra. Concerning the symphony, he wrote his wife that "it hardly will be published quickly because of the large amount of work on *Igor*, which progresses slowly"[45].

Twelve days after this letter, on the morning of February 15th [February 27th, N. S. — T.] Borodin composed the finale [fourth movement. — T.] of the Third Symphony, but he did not succeed in writing it down. In the evening he went to a fancy-dress dancing party, organized by the professors of the Medical-Surgical Academy.

At 11:40 P. M., in the midst of the most natural merriment, Borodin suddenly approached the wall, leaned against it and there, without a moan, fell dead on the floor.[46]

With extreme sorrow and grief all the prominent Russian society attended the funeral of Aleksandr Porfir'evich Borodin, the great composer, remarkable chemist, and tireless advocate of women's education. According to Stassov,

> All his admirers, who were very numerous, all the Medical Academy, all the women medical doctors of ten graduations who were located in [St.] Petersburg, attended the funeral. The young people carried him in their arms to the grave, men and women; a whole forest of garlands attended his body[47].

Members of the Russian Physico-Chemical Society, noting the severe loss for Russian science, wrote in the *Journal of the Russian Physico-Chemical Society*:

> A. P. Borodin belonged to the founding members of our society, thus being a member of the society since 1868. The sympathetic image of the deceased is no doubt memorable to the majority of the members of the section[48].

Borodin was buried next to his friend Mussorgsky not far from the grave of Dargomyzhskiĭ and Aleksandr Nikolaevich Serov.

Rimsky-Korsakov, Aleksandr Profir'evich's closest friend and comrade, recalling the sad loss, wrote in his Letopis' moeĭ muzykal'noĭ zhizni (Chronicle of My Musical Life):

> Early in the morning, at an unaccustomed hour, on February 16, 1887, I was astonished by a visit from V. V. Stassov at our house. V. V. was beside himself. "Do you know," he said with agitation, "Borodin is dead." Borodin had died late the evening before, suddenly, instantaneously. Gay and animated, among guests gathered at his house, he had fallen stark dead, in the very act of talking to someone . . . I shall not say what a blow this death was to myself and all his other intimates. Immediately the question came up: what was to be done with the unfinished opera "Prince Igor" and his other unpublished and unfinished compositions? Together with Stassov we forthwith went to his apartment and fetched all his musical manuscripts to my house.
>
> After Aleksandr Porfir'evich had been buried at the cemetery of the Nevskiĭ Monastery, Glazunov and I together sorted all the manuscripts. We decided to finish, orchestrate, and set in order all that had been left behind by A. P., as well as prepare it for publication on which M. P. Beliaev had resolved[49].

Beliaev[50], an ardent lover of Russian music, wanting in every way to promote the diffusion and popularization of the great musical productions of Russian composers, published the musical compositions of Glazunov, Borodin, Rimsky-Korsakov, Liadov, Cui, and many other Russian composers in beautiful editions.

In 1884 Beliaev had established awards for new compositions of Russian composers, which had been received by Borodin, Balakirev, Rimsky-Korsakov, Tchaikovsky, and Liadov in the very first year. In 1887 the Beliaev Award was granted for the erection of a monument to Aleksandr Profir'evich Borodin.

Two years later (February 15, 1889) Borodin's friends and admirers erected a beautiful memorial at his grave. The monument portrayed a "golden musical page" left in our history by Borodin. On this page several themes from Borodin's most important musical works were inscribed in mosaic on a gold background: the first theme from his Second Symphony, the chorus of Polovtsian women (from the second act of the opera *Prince Igor*), the beginning of The Song of the Dark Forest, the first theme of the scherzo from the Third Symphony, and the beginning of the symphonic poem *In Central Asia*.

On a projection in front of the golden page stands a bronze bust of Borodin. Under the bust are two Russian musical instruments, a gusli[51] (psaltery) and a gudok[52] (rebec). Above the golden page is the inscription in granite *Borodinu* (To Borodin).

The inscription is shaded by two branches, which ascend to the right and to the left. Below are the years of Borodin's birth and death: 1834–1887[53].

Around the monument is a grating of forged iron. In the middle of it the monogram A. B. is represented, surrounded by three garlands one inside the other. Of these one is of chemical formulas,

$$C_7H_{14}O \mid C_{10}H_{18}O \mid C_{10}H_{22}O \mid C_{10}H_{20}O \mid C_{10}H_{20}O_2 \mid$$

$$CH_3-\overset{\overset{OH}{|}}{C}H-CII_2CHO\ ,$$[54]

chemical compounds discovered by Borodin. The other garland is of Borodin's musical themes, and the third garland is of laurel.

The monument to Borodin was created by the architect I. P. Ropett[55].

On April 23, 1890 Borodin's opera *Prince Igor* was performed for the first time on the stage of the Mariinsky Opera Theater[56] in St. Petersburg and was an enormous success.

In addition to the opera Glazunov restored two parts of Borodin's Third Symphony[57]. It was performed for the first time in October, 1887, and somewhat later it was published by M. P. Beliaev, who also issued a luxurious edition of the orchestral score and piano score of *Prince Igor*.

Fig. 14. Memorial on the grave of A. P. Borodin at the cemetery of the Aleksandr Nevskiĭ Monastery (Leningrad)

In the history of Russian science Borodin is recognized as an outstanding scholar, an investigator with wide and versatile interests, a bold experimenter, and a remarkable teacher. He trained an entire galaxy of Russian chemists and transmitted to them the glorious tradition of native science, to which he had made an earnest contribution.

Borodin's chemical works, original in content and absolutely new in conception, played a great role in the growth of organic and physiological chemistry.

Borodin's investigations on the fluorination of organic compounds, the many years of research on the condensation of aldehydes, and the works on the study of esters and fatty acids are excellent examples of his chemical activity. It should be mentioned that his chemical findings were not carried through as far as those of some foreign scholars who, acquainted with his original works, adopted his research ideas and continued work in this field on a scale wider than that which the author of the idea was able to accomplish under unfavorable conditions. Borodin's collisions with Schützenberger, Kekulé, and Wurtz are well known. Solutions of scientific problems in organic chemistry which interested Borodin involved great difficulty from an experimental as well as from a theoretical aspect. If one remembers what conditions and with what laboratory technique Borodin carried out research on the condensation of aldehydes, in particular the aldol condensation, and if one appraises his work on the study of the structure of complex organic compounds (amarine and others), then it becomes clear that Borodin occupies an outstanding place among the remarkable Russian scholar-chemists who played a large role in the growth of world chemical science.

In the history of Russian music Borodin is recognized as one of the most brilliant of its representatives. A direct successor of Glinka, a comrade of Balakirev, Mussorgsky, Rimsky-Korsakov, and Cui, he created compositions of genius, characterized by a national trait of Russian musical art and at the same time by genuine originality, peculiar to his genius.

Among these compositions the opera *Prince Igor* is outstanding. In this opera the versatility of Borodin's creative genius became apparent to the greatest extent.

Borodin's creativity in composition was one of the most significant stages in the history of the growth of Russian classical music. Continuing Glinka and Dargomyzhskiĭ's struggle for original, national musical art, Borodin enriched it with new forms, extended and developed the patriotic tradition of Russian classical music, created new means of expression, and boldly used them, appearing as a genuine innovator in art.

The influence of Borodin's music and achievements has beneficially affected even his contemporaries. Rimsky-Korsakov himself recognized that the idea for his composition *Scheherazade*[58] occurred to him when he was completing and attentively studying Borodin's compositions. It is quite evident that the coloring of this splendid production in one way or another is associated with Borodin's interpretation of oriental music. Borodin's creative works exerted a noticeable influence on his young contemporaries, students of Rimsky-Korsakov, Glazunov, and Liadov. For the source of such works of Liadov as his popular Russian epic *Pro starinu* (From Days of Old)[59], one should undoubtedly look to Borodin's Bogatyr Symphony, whose influence had already affected Liadov in his earliest productions. Borodin's influence also beneficially influenced the productions of many Soviet composers.

Beyond doubt, the profound study and creative mastery of the classical musical legacy of Borodin helps Soviet composers to develop successfully the tradition of the Russian musical school and to reach the heights of socialist realism.

The Soviet people are passionately interested in the history of their native land and the legacy left by their famous representatives, who are outstanding figures of science and the arts. Their inspired creative works and bold, daring scientific acts invoke the legitimate pride of our great nation.

This legacy has a high price; our nation remembers with pride the investment in Russian culture which was made by one of its outstanding sons — Aleksandr Porfir'evich Borodin.

Footnotes to Chapter 12

1 Vasiliĭ Yakovlevich Danilevskiĭ (1852–1934), a physiologist, was Professor of Physiology at Kharkov University from 1883–1911. — A. [See GSE, 1975, Vol. 7, p. 100. — T.]

2 Mikhail Egrafovich Saltykov (1826–1889) was a prominent Russian writer and scientist, who wrote under the pen name of N. Shchedrin (See GSE, 1979, Vol. 22, pp. 567–568). — T.

3 Letter of V. Ya. Danilevskiĭ to A. M. Butlerov, 1885, Archives of the Academy of Sciences of the USSR, Stock 22, Inventory 2, No. 55, published here for the first time. — A.

4 Sbornik postanovleniĭ po Ministerstvu narodnogo prosveshcheniia (Collection of Decisions Concerning the Ministry of Public Education), Vol. 10, St. Petersburg, 1894, p. 881. — A.

5 According to A. P. Dianin, when he and Borodin moved the equipment of the chemical laboratory back to the Academy of Medicine, Borodin burst into tears. — T.

6 Borodin: Letters, III, p. 252. — A.

7 Ibid., p. 203. — A.

8 Borodin's wife's uncle by marriage, a typical example of an official. — T.

9 Borodin brought up several orphans. — A. [Elizaveta (Lisa) Gavrilovna Balaneva, Elena A. Guseva, and Gania Litvinenko were adopted by the Borodins and lived with them. The first-named married Aleksandr Pavlovich Dianin, Borodin's student and successor, and became the mother of Sergeĭ Aleksandrovich Dianin. — T.]

10 V. V. Stassov: Borodin, p. 179. — A.

11 V. V. Stassov: Letters, p. 9. — A.

12 The Russian Physico-Chemical Society was created in 1878 by merging the Russian Physical Society (founded in 1872) and the Russian Chemical Society (founded in 1868). — A. [For the name changes of this society see footnote 3, Complete List of Borodin's Chemical Works. — T.]

13 Reported by Georg Wagner in Ber. *12*: 693 (1879). — T.

14 $p\text{-}NO_2C_6H_4CH_2COC_6H_5 \xrightarrow{Sn,\ HCl} p\text{-}NH_2C_6H_4CH_2COC_6H_5$. — T.

15 A long-chain fatty acid, used in biochemical research; triacontanoic acid, $CH_3(CH_2)_{28}COOH$. — T.

16 Reported by Georg Wagner: In: *Ber. 12*: 696–697 (1879). — T.

17 Reported by Georg Wagner: In: *Ber. 13*: 2403 (1880). See also *Zh. Russ. khim. 13* (I): 23 (1881); abstracted by Jawein as Ueber Dinitrodesoxybenzoine und deren Oxydation durch Chromsäure. *Ber. 14*: 2067–2068 (1881). — T.

18 A city in the Georgian Soviet Socialist Republic. — T.

19 *Zh. Russ. khim. 13* (I): 32 (1881); abstracted by Jawein as Ein Doppelsalz des Chlorwasserstoff-Chinins mit dem Chlorwasserstoff-Harnstoffe. In: *Ber. 14*: 2069 (1881). The three works in this paragraph by Bronnikov, Golubev, and Drygin are reported in *Bulletin de la Société Chimique de France 35*: 560–561 (1881). — T.

20 See Appendix X. — A.

21 Letter of M. Yu. Goldstein to A. M. Butlerov, May 15, 1885, Archives of the Academy of Sciences of the USSR, Stock 22, Inventory 2, No. 47, published here for the first time. — A.

22 Borodin: Letters, III, p. 137. — A.

[23] M. M. Kurbanov: Otryvki iz vospominaniĭ (Fragments of Reminiscences) [hereafter abbreviated as Fragments], Biriuch, Petrograd, 1920, p. 165.
[24] N. A. Rimsky-Korsakov: My Musical Life, p. 175. — A. [English version, pp. 176–177. — T.]
[25] Glinka's opera, Ruslan and Liudmila. — T.
[26] This aria opens Act IV (No. 25). — T.
[27] V. V. Stassov: Letters, pp. 6–7. — A.
[28] Arthur Nikisch (1855–1922), born in Hungary, one of the finest conductors of the nineteenth century, was Principal Conductor of the Leipzig Opera at the time. — T.
[29] Symphony No. 2, Op. 9 was written during the summer of 1868, but Rimsky-Korsakov revised it three times thereafter. — T.
[30] Borodin: Letters, II, p. 182. — A.
[31] As Borodin in jest called Liszt in order to emphasize his charm. — A.
[32] Borodin: Letters, III, p. 163. — A.
[33] Letter of A. P. Borodin to A. P. Dianin, August 3, 1881, published here for the first time. — A.
[34] Dedicated to Borodin's wife and published by Beliaev in 1890. — T.
[35] In D Major, published by Beliaev in 1888. — T.
[36] This 7-movement programmatic suite was published by Bessel in 1885. Vasiliĭ Vasil'evich Bessel (1843–1907), a Russian music publisher, published works by all the prominent Russian composers, including Tchaikovsky, Dargomyzhskiĭ, Anton Rubinstein, and, of course, Borodin and the Mighty Little Group (See Geoffrey Norris: Vasily Vasil'yevich Bessel. In: Grove, Vol. 2, p. 160). — T.
[37] Countess Louise of Mercy-Argenteau, to whom the Petite Suite and other works by Borodin were dedicated, was a devoted admirer of and propagandist for Borodin's works. — T.
[38] M. M. Kurbanov: Fragments, p. 163. — A.
[39] Scherzo (Allegro vivace) in A Flat Major, dedicated to the Belgian pianist, composer, and conductor Théodore Jadoul and published by Bessel in 1885. Aleksandr Konstantinovich Glazunov orchestrated it as the finale (No. 7) for the Petite Suite. — T.
[40] V. V. Stassov: Borodin, p. 245. — A.
[41] Ballad for voice and piano in F Major, words by Borodin, dedicated to A. E. Makovskaia, published by Bessel in 1873. — T.
[42] A fairy tale for voice and piano, words by Borodin, dedicated to Rimsky-Korsakov, published by Jurgenson in 1870. Pëtr Ivanovich Jurgenson (1836–1904) founded the largest musical publishing firm in Russia, which had a branch in Leipzig. The principal publisher of the compositions of Tchaikovsky and many other Russian composers, the firm was nationalized in 1918 and became the music section of the Soviet State Publishing House (See Geoffrey Norris: Pyotr Ivanovich Jürgenson [Yurgenson]. In: Grove, Vol. 9, p. 753. — T.
[43] V. V. Stassov: Borodin, p. 246. — A.
[44] The Symphony No. 3, in A Minor was left unfinished on Borodin's death. The first (Moderato assai) and second (Vivo) movements were edited by Glazunov and in 1888 were published by Beliaev. The second movement (Scherzo, D Major) was also adapted for string quartet as Les Vendredis; it was probably composed for one of Beliaev's Fridays. (See footnote 50, this chapter. — T.
[45] V. V. Stassov: Borodin, p. 269. — A.
[46] The autopsy, carried out in the same room in which the last dance had been held, disclosed that the cause of death was a burst artery in Borodin's heart. The artery wall had become so fragile that it could not support the pressure of the blood. Also, when Borodin fell, he struck his temple against the corner of the stove where he had been standing, causing a minor cerebral hemorrhage. Borodin's wife survived him for only a few months; she died on June 28, 1887. — T.
[47] V. V. Stassov: Biography, p. 167. — A.
[48] Zhurnal Russkogo fiziko-khimicheskogo obshchestva (Journal of the Russian Physico-Chemical Society), 1888, No. I, p. 1. — A.
[49] N. A. Rimsky-Korsakov: My Musical Life, pp. 230–231. — A. [English version, p. 238. — T.]
[50] Mitrofan Petrovich Beliaev (1836–1904) published many of Borodin's compositions. (See GSE, 1973, Vol. 3, p. 140, Rosa Newmarch and Geoffrey Norris: Mitrofan Petrovich Byelyaiev. In: Grove, Vol. 2, p. 458, and V. V. Stassov: Mitrofan Petrovich Beliaev, Gosudarstvennoe muzykalnoe muzykal'noe izdatel'stvo, Moscow, 1954. Borodin had frequently attended the chamber music evenings held at Beliaev's house known as "the Beliaev Fridays." As Russian composers gathered

around Balakirev in the 1860's (the Mighty Little Group), a similar group gathered around Beliaev in the 1880's. For a comparison of the two groups by Rimsky-Korsakov, who was a member of both groups, see his book *My Musical Life*, 2nd, rev. ed., Alfred A. Knopf, New York, 1928, pp. 239–244. — T.

[51] An ancient musical instrument consiting of a flat sounding box with numerous strings, which are plucked with the fingers or with a plectrum. — T.

[52] See footnote 3, Chap. 11. — T.

[53] Concerning the dates of Borodin's birth see p. 9 and footnote 2, Chap. 1. — A.

[54] The six compounds depicted are enanthaldehyde (see pp. 68–70), a condensation product of valeraldehyde (see pp. 68–70, 83), "isocapric" alcohol (see p. 62), "isocapric" aldehyde (see p. 63) "isocapric" acid (see p. 62, 63), and aldol (2-hydroxybutyraldehyde) (see p. 70), respectively. — T.

[55] Ivan Pavlovich Ropett (1845–1908). — T.

[56] See footnote 15, Chap. II. — T.

[57] See footnote 44, this Chapter. — T.

[58] This symphonic suite, Op. 35, is Rimsky-Korsakov's best known work. It is based on episodes from *The Thousand and One Nights* and was completed in 1888, the year after Borodin's death. Scheherazade is the sultana who tells the sultan a new and fascinating story each night for a thousand and one nights, never revealing each story's conclusion until the following evening. The sultan, unable to contain his curiosity about the outcome of these wondrous tales, delays her execution from day to day. Like Borodin, Rimsky-Korsakov, who was a widely traveled naval officer, favored "oriental" themes for his compositions. — T.

[59] Ballad, Op. 21. — T.

Complete List of Borodin's Chemical Works

1858

1. Recherches sur la constitution chimique de l'hydrobenzamide et de l'amarine (Researches on the Chemical Constitution of Hydrobenzamide and Amarine) (Read on March 5, 1858), *Bulletin de la classe physico-mathématique de l'Académie des Sciences de St. Pétersbourg* [hereafter abbreviated as *Bull. Acad. Sci., 17*], Nos. 1, 2, 3, notes, 38–46 (1858/1859) [Ueber die Constitution des Hydrobenzamids und des Amarins, *Justus Liebigs Annalen der Chemie* (hereafter abbreviated as *Ann.*), *110*, 78–85 (1859). — T.]
2. Ob' analogii mysh'iakovoĭ kisloty s fosfornoiu v khimicheskom i toksikologicheskom otnosheniiakh (On the Analogy Between Arsenic Acid and Phosphoric Acid in Chemical and Toxicological Behavior), Dissertation, [Tipografiia Koroleva,] St. Petersburg, 1858. Dissertation for the Degree of Doctor of Medicine. Defense [of the dissertation] took place on May 3, 1858 at the Medical-Surgical Academy; translated into English by Harry Rosenberg, The Australian National University, Canberra, 1987.
3. Ueber die [Ein]wirkung des Jodäthyls auf Benzoylanilid [O deĭstvii iodistogo etila na benzoilanilid] (On the Action of Ethyl Iodide on Benzoylanilide). (Read on November 26, 1858), *Bull. Acad. Sci., 17*, No. 26, notes, 408–411 (1859) [*Journal für praktische Chemie* (hereafter abbreviated as *J. prakt. Chem.*), *77*, 19–22 (1859); *Ann.*, *111*, 254–256 (1859). — T.]

1859

4. Soligalicheskie soleno-mineral'nye vody (Soligalich Saline-Mineral Waters), *Moskovskie vedomosti (Moscow News)*, *130*, 972–974 (June 3, 1859); also published by Tipografiia Universiteta (University Printers), Moscow, 1859, 18 pp., 3 tables.

1860

5. Ueber die Einwirkung des Jodäthyls auf Benzidin [O deĭstvii iodistogo etila na benzidin] (On the Action of Ethyl Iodide on Benzidine), *Zeitschrift für Chemie und Pharmazie* [hereafter abbreviated as *Z. Chem. Pharm.*], *3*, 533–536 (1860).
6. Über einige Derivate des Benzidins [O nekotorykh proizvodnykh benzidina] (On Several Derivatives of Benzidine), *Z. Chem. Pharm.*, *3*, 641–643 (1860).
7. Sur les derivés monobromes des acides valérique et butyrique (On the Monobromine Derivatives of Valeric and Butyric Acids) (Read on November 23, 1860), *Société Chimique de Paris, Bulletin des Séances de 1858–1860*, 252–254 (1861) [Ueber die Monobrombaldriansäure und Monobrombuttersäure (O bromovalerianovoĭ i bromomaslianoĭ kislote), *Z. Chem. Pharm.*, *4*, 5–7 (1861); Ueber Bromvaleriansäure und Brombuttersäure, *Ann.*, *119*, 121–123 (1861); *J. prakt. Chem.*, *84*, 474–475 (1861). — T.]

1861

8. Ueber die Wirkung des Zinkäthyl's [sic] auf zusammengesetzte Aether [(O deĭstvii slozhnykh efirov na tsinketil)] (On the Action of Zinc Ethyl on Complex Esters), *Z. Chem. Pharm.*, *4*, 8–12 (1861). [also see Fatti per servire alla storia di benzile (Data for the History of Benzil), *Il Nuovo Cimento*, *15*, 314–315 (1862) (No. 11 below). — T.]

1862

9. Beitrag zur Geschichte des Benzils [(Contribution to the History of Benzil)], *Z. Chem. Pharm.*, *5*, 580–581 (1862).

10. [Fatti per servire alla storia de' fluoruri (O ftoristom benzoile, k istorii ftoristykh soedinenii) (Data for the History of Fluorides) — T.], *Il Nuovo Cimento*, *15*, 305[-314] (1862) [Faits pour servir à l'histoire des fluorures et préparation du fluorure de benzoyle, *Comptes rendus des séances de l'Académie des Sciences (Paris)*, *55* (13), 553–556 (1862); Contributions to the History of Fluorides, *Chemical News*, *6*, 267–268 (November 29, 1862); Zur Geschichte der Fluorverbindungen und über das Fluorbenzoyl, *Ann.*, *126*, 58–62 (1863). — T.][1]
11. [Fatti per servire alla storia di benzile (K istorii benzila)] (Data for the History of Benzil) — T.], *Il Nuovo Cimento*, *15*, 314[-315] (1862); [Ueber die Einwirkung des Benzils auf Natrium-Amylat, *Ann.* *126*, 372–373 (1863). — *T.*][1]
12. [Sull'azione dello zincoetile sul cloroiodoforme (O deĭstvii tsinketila na khloriodoform) — T.] (On the Action of Zinc Ethyl on Chloroiodoform), *Il Nuovo Cimento*, *15*, 316 [—319] (1862); extract published in *Répertoire de chimie pure*, *4*, 431—432 (1862); Ueber die Einwirkung des Zinkäthyls auf das Chloroiodoform, *Ann.*, *126*, 239–241 (1863). — T.][1]

1863–1864

13. Referat ob uspekhakh farmatsii v 1861 godu (Report on Progress in Pharmacy in 1861), *Voenno-meditsinskiĭ zhurnal (Military-Medical Journal)*, *88* (10), Sec. 6, Misc., 220–234 (October, 1863); *88* (11), Sec. 6, Misc., 289–306 (November, 1863); *88* (12), Sec. 6, Misc., 371–403 (December, 1863); *90* (6), Sec. 6, Misc., 145–160 (June, 1864).
14. [O deĭstvii natriia na valerianovyl al'degid] (On the Action of Sodium on Valeraldehyde) (Read on May 20, 1864), *Bull. Acad. Sci.*, *7*, 463–474 (1864); Ueber die Einwirkung des Natriums auf Valeraldehyd, [*J. prakt. Chem.*, *93*, 413–425 (1864). — T.]

1865

15. Referat po farmatsii za 1862 god (Report on Pharmacy for 1862), *Voenno-meditsinskiĭ zhurnal (Military-Medical Journal)*, *92* (1), Sec. 6, Misc., 3–33 (January, 1865); *92* (2), Sec. 6, Misc., 93–122 (February, 1865); *92* (3), Sec. 6, Misc., 163–179 (March, 1865).

1866–1869

Analiticheskaia Khimiia (Analytical Chemistry), edited by D. Mendeleev, Kolichestvennyĭ analiz (Quantitative Analysis), Vols. 1–3, Obshchestvennaia pol'za (Public Benefit), St. Petersburg, 1866–1869.

In the preface to the first volume Mendeleev wrote: "The third volume includes descriptions of those most important applied analyses which are carried out by special methods . . . In this issue Professor of the Medical-Surgical Academy A. P. Borodin will write a section on medical (clinical) analysis." The third volume appeared in 1869 and was devoted to technical and agricultural analysis. Borodin was absent from the list of authors of articles in this volume, but the introduction stated: "A description of methods of medical-chemical research was supposed to be included in this issue, but the vastness and specialty of this subject forces us to allot a special volume for it."[2]

1867

16. Issledovanie bukharskogo opiia (Research on Bukhara Opium), *Protokoly obshchestva russkikh vracheĭ za 1867 god (Transactions of the Society of Russian Physicians for 1867)*, 221.

1868

17. O proizvodnykh valer'ianovogo al'degida (On Derivatives of Valeraldehyde), *Trudy Pervogo s'ezda russkikh estestvoispytateleĭ v S.-Peterburge, proiskhodivshego s 28 dekabria 1867 g. po 4*

---

[1] Borodin carried out these works while staying in Pisa during the winter of 1861–1862. — A.

[2] No further traces of this special volume were found; evidently this publication did not appear, and Borodin's work, if it was written, remained unpublished. Two of Borodin's letters to Mendeleev exist in which this question is partially elucidated. — A.

yanvaria 1868 g. (Proceedings of the First Meeting of Russian Naturalists at St. Petersburg, Held from December 28, 1867 to January 4, 1868), St. Petersburg, *4*, 13 (1868).

1869

18. O produktakh deĭstviia parov broma na serebrianye soli kislot maslianoĭ i valerianovoĭ (On the Products of the Action of Bromine Vapors on Silver Salts of Butyric and Valeric Acids), *Zhurnal Russkogo khimicheskogo obshchestva* [hereafter abbreviated as *Zh. Russ. khim.*] *(Journal of the Russian Chemical Society)*,[3] *1* (1), 31–32 (1869); recorded note of a communication made at a meeting of the society on January 9, 1869; [Ueber die Einwirkung von Brom auf buttersaures und valeriansaures Silber, *Z. Chem. Pharm.*, *12*, 342 (1869). — T.]
19. Izokaprinovaia kislota, ee al'degidy i soli (Isocapric Acid, Its Aldehyde and Salts), *Trudy Vtorogo s'ezda russkikh estestvoispytateleĭ v Moskve, proiskhodivshego s 20 po 30 avgusta 1869 (Proceedings of the Second Meeting of Russian Naturalists at Moscow, Held on August 20 to 30, 1869)*, Moscow, *1* (Section on Chemistry), 6 (1870); untitled recorded note of a communication made at a meeting on August 22, [1869]; also [published in] *Zh. Russ. khim.*, *1* (8–9), 225 (1869).
20. Produkty uplotneniia al'degidov (Condensation Products of Aldehydes), *Zh. Russ. khim.*, *1* (8–9), 214 (1869); untitled recorded note of a communication made at a meeting of the society on October 2, 1869 (A. P. Dianin called this work O deĭstvii vysokoĭ temperatury na enantol i valeral'degid (On the Action of High Temperature on Enanthol and Valeraldehyde).) [Abstracted by Victor von Richter in *Berichte der Deutschen Chemischen Gesellschaft zu Berlin* (hereafter abbreviated as *Ber.*), *3*, 552–553 (1870). — T.]

1870–1871

Editor and publisher, together with P[ëtr Petrovich?] Khlebnikov, of the journal *Znanie (Knowledge)*.

1870

21. [O proizvodnykh odnogo izokaprinovogo riada] Über die Derivate einer Isokaprinreihe (On Derivatives of an Isocapric Series) (Read on January 27, 1870), *Bull. Acad. Sci.*, *14*, 535–542 (1870).
22. Produkty uplotneniia valerianovogo al'degida (Condensation Products of Valeraldehyde), *Zh. Russ. khim.*, *2* (4), 90–91 (1870); untitled recorded note of a communication made at a meeting of the society on March 5, 1870. [Abstracted by Victor von Richter, in *Ber.*, *3*, 423 (1870). — T.]

1872

23. O deĭstvii natriia na valeral'degid (On the Action of Sodium on Valeraldehyde), *Zh. Russ. khim.*, *4* (6), 207–208 (1872); untitled recorded note of a communication made at a meeting of the society on May 4, 1872. [Abstracted by Victor von Richter in *Ber.* *5*, 480–481 (1872). — T.][4]

---

[3] Through the years this journal has undergone a number of name changes. It was called *Zhurnal Russkogo khimicheskogo obshchestva* from Vol. 1, No. 1 (1869) through Vol. 4, No. 9 (1872); *Zhurnal Russkogo khimicheskogo obshchestva i fizicheskogo obshchestva (Journal of the Russian Chemical Society and of the Physical Society)* from Vol. 5, No. 1 (1873) through Vol. 10, No. 9 (1878); and *Zhurnal Russkogo fiziko-khimicheskogo obshchestva (Journal of the Russian Physico-Chemical Society)* from Vol. 11, No. 1 (1879) through Vol. 38, No. 9 (1906). In 1907 the publication split into two parts – (1) *Zhurnal Russkogo fiziko-khimicheskogo obshchestva, Chast' khimicheskaia (Journal of the Russian Physico-Chemical Society, Chemical Part)* from Vol. 39, No. 1 (1907) through Vol. 62, No. 10 (1930), after which it was entitled *Zhurnal obshcheĭ khimii (Journal of General Chemistry)* beginning with 1931 (Vol. 1, No. 1); and (2) *Zhurnal Russkogo fiziko-khimicheskogo obshchestva, Chast' fizicheskaia (Journal of the Russian Physico-Chemical Society, Physical Part)* from Vol. 39, No. 1 (1907) through Vol. 62, No. 6 (1930), after which it was entitled *Zhurnal eksperimental'noĭ teoreticheskoĭ fiziki (Journal of Experimental and Theoretical Physics)* beginning with 1931 (Vol. 1, No. 1). We have consistently abbreviated the journal here as *Zh. Russ. khim.* — T.

[4] See page 124.

24. O deĭstvii natriia na enantol (On the Action of Sodium on Enanthol), *Zh. Russ. khim.*, *4* (6), 208 (1872); untitled recorded note of a communication made at a meeting of the society on May 4, 1872. [Abstracted by Victor von Richter in *Ber.*, *5*, 481 (1872). — T.][4]
25. O poluchenii produkta uplotneniia obyknovennogo al'degida (On Condensation Products of Common Aldehyde), *Zh. Russ. khim.*, *4* (6), 209 (1872); untitled recorded note of a communication made at a meeting of the society on May 4, 1872. [Abstracted by Victor von Richter in *Ber.*, *5*, 481–482 (1872). — T.][4]

1873

26. Über einen neuen Abkömmling des Valerals [O novykh proizvodnykh valeral'degida] (On a New Derivative of Valeral) (Received on July 18), *Ber.*, *6*, 982–985 (July–December, 1873). (A. P. Dianin called this work O gidrate polimera valeral'degida (On the Hydrate of a Polymer of Valeraldehyde).) [Abstracted by H. E. A. (Henry Edward Armstrong) in *Journal of the Chemical Society, London* (hereafter abbreviated as *J. Chem. Soc.*), *27*, 145 (1874). — T.]
27. Issledovanie polimera valeral'degida (Research on the Polymer of Valeraldehyde), *Trudy Chetvertogo s'ezda russkikh estestvoispytateleĭ v Kazani, proiskhodiveshego s 20 po 30 avgusta 1873 g. Po otdeleniiam khimii, mineralogii, geologii i paleontologii . . .* [hereafter abbreviated as *Trudy Chet. Kazani*] *(Proceedings of the Fourth Meeting of Russian Naturalists at Kazan, Held from August 20 to 30, 1873, at the Sections on Chemistry, Mineralogy, Geology, and Paleontology . . .)*, Kazan, 5–6 (1875); untitled recorded note of a communication made at a session of the Chemical Section of the meeting on August 23, 1873; also [published in] *Zhurnal Khimicheskogo i Fizicheskogo obshchestva* [hereafter abbreviated as *Zh. Khim. Fiz.*], *5* (8), 386, 410 (1873).
28. Gidramidy i izmernye s nimi shchelochi (Hydramides and Alkalis Isomeric with Them), *Trudy Chet. Kazani*, Kazan, 6–7 (1875); untitled recorded note of a communication made at a session of the Chemical Section of the meeting on August 23, 1873; also [published in] *Zh. Khim. Fiz.*, *5* (8), 387, 410 (1873); recorded note of a communication made at the fourth meeting.[5]
29. O suktsinildibenzoine (On Succinyldibenzoin), *Trudy Chet. Kazani*, Kazan, 7 (1875); untitled recorded note of a communication made at a session of the Chemical Section of the meeting on August 23, 1873; also [published in] *Zh. Khim. Fiz.*, *5* (8), 388 (1873); recorded note of a communication made at the fourth meeting.
30. O deĭstvii ammiaka na kuminol (On the Action of Ammonia on Cuminol), *Trudy Chet. Kazani*, Kazan, 6 (1875); untitled recorded note of a communication made at a session of the Chemical Section at the meeting on August 23, 1873; also [published in] *Zh. Khim. Fiz.*, *5* (8), 389 (1873); recorded note of a communication made at the fourth meeting.[5]

1875

31. Über Nitrosoamarin (On Nitrosoamarine) (Received on July 10), *Ber.*, *8*, 933–936 (1875); also O nitrosoamarine, *Zh. Khim. Fiz.*, *7* (8), 285–286 (1875); recorded note of a communication made at a meeting of the society on October 2, 1875. [Abstracted by G. T. A. (G. T. Atkinson) as Nitrosoamarine in *J. Chem. Soc.*, *29*, 269–270 (1876). — T.]

1875–1876

32. O novom sposobe opredeleniia azota v moche (On a New Method for the Determination of Nitrogen in Urine), *Protokoly Obshchestva russkikh vracheĭ v S.-Peterburge (Transactions of the Society of Russian Physicians at St. Petersburg)*, 278 (1875–1876).
33. O novom sposobe kolichestvennogo opredeleniia mocheviny (On a New Method for the Quantitative Determination of Urea), *Zh. Khim. Fiz.*, *8* (5), Chemical Section, 145 (1876); untitled

---

[4] Von Richter gives the date of the St. Petersburg meeting as April 6. All three papers are abstracted by C. S. (Carl Schorlemmer) as Condensation-products of Aldehydes, *J. Chem. Soc.*, *26*, 58 (1873). — T.

[5] Abstracted by Victor von Richter as Constitution des Hydrobenzamids und dessen Umwandlung in Amarin in *Ber.*, *6*, 1253–1254 (1873) and by C. S. (Carl Schorlemmer) as Constitution of Hydrobenzamide and Its Conversion into Amarine in *J. Chem. Soc.*, *27*, 273 (1874). — T.

recorded note of a communication made at a meeting of the society on March 25, 1876 with a short description of the apparatus invented by Borodin. [Abstracted by Georg Wagner (Egor Egorovich Vagner) in *Ber.*, *9*, 1029 (1876). — T.]

1877

34. O nekotorykh izmeneniiakh pribora dlia opredeleniia mocheviny v moche (On Some Changes in the Apparatus for the Determination of Urea in Urine), *Zh. Khim. Fiz.*, *9* (6), Chemical Section, 240 (1877); untitled recorded note of a communication made at a meeting of the society on May 5, 1877. [Mentioned by Georg Wagner (Egor Egorovich Vagner) in *Ber.*, *10*, 1105 (1877). — T.]

1878

35. Azometricheskiĭ sposob opredeleniia mocheviny s tablitsami: (1) vesovykh kolichestv mocheviny; (2) vesovykh kolichestv azota (A Method for the Determination of Urea by the Measurement of Nitrogen with Tables: (1) Gravimetric Amounts of Urea; (2) Gravimetric Amounts of Nitrogen), in D. Koshlakov's book, Analiz mochi. Klinicheskoe rukovodstvo dlia studentov i vracheĭ (The Analysis of Urine. A Chemical Manual for Students and Physicians), St. Petersburg, 1878, p. 6; also 2nd edition, St. Petersburg, 1880, pp. 83–116. The title of this work is missing. Koshlakov's book states [merely]: "Prof. Borodin's Method." The fact that the description of the method and the tables affixed to the method belong to Borodin is stated in the introduction to Koshlakov's book and in Borodin's work: Uproshchennyĭ azometricheskii sposŏb . . . (A Simplified Azometric Method . . .) (See Work No. 42), in which Borodin points out that "the detailed description is made by me only in Russian and is located in D. I. Koshlakov's paper . . .".
36. O dezinfektsii i dezinfektsionnykh sredstvakh (On Disinfection and Disinfecting Agents), *Zdorov'e (Health)*, *85*, 142–143 (April 15, 1878). The work was written together with A. P. Dobroslavin. Borodin and Dobroslavin were among the members of a Commission for the Investigation of Measures against the Propagation of Typhus Epidemics in St. Petersburg; this article is a note of a report made in March, 1878 at the combined meeting of the commission with the council and sections of the Russian Society of the Preservation of Public Health.

1880

37. Nikolaĭ Nikolaevich Zinin. Vospominaniia o nem i biograficheskiĭ ocherk (Nikolaĭ Nikolaevich Zinin. Reminiscences of him and a Biographical Sketch), *Zh. Khim. Fiz.*, *12* (5), Chemical Section, 215–254 (1880). This work was written together with A. M. Butlerov; also read at a meeting of the Physical-Mathematical Section of the Academy of Sciences on May 13, 1880; *Zapiski Akademii Nauk (Proceedings of the Academy of Sciences)*, *37*, Book 1, 1–46 (November, 1880).

1883

38. Narodnyĭ chaĭ (plitki Ponomareva) (Native Tea (Ponomarev Bricks)), *Zdorov'e (Health)*, *9*, 3–5 (February 27, 1883). An analysis of tea and evaluation of its grade. Borodin showed that cheap tea in bricks is of the best quality since it does not contain harmful substitutes.
39. Analiz plitochnogo chaia (fabrika Ponomareva v Khan'kou) (The Analysis of Brick Tea (Ponomarev factory at Khan'ka))[6], *Nauchno-sanitarnye novosti (Scientific Sanitary News)*, No. 3, 1–5 (March, 1883). A large part of the article is devoted to the defense of a method for the analysis of tea, used by Prof. Markovnikov, unjustly ascribed to A. Fogel. Borodin carried out both methods and pointed out the originality and the better qzality of Markovnikov's method compared to Fogel's method.

1884

40. Ob otnoshenii perekisi vodoroda k nizshim organizmam i znachenii vodnogo rastvora ozonirui-ushchikh masel dlia dezinfektsii (On the Relationship of Hydrogen Peroxide toward Lower

---

[6] Hankow, a large urban area and river port in Wu-han Municipality, Hupeh Province, central China, was one of the first Chinese cities opened to foreign trade (1861). — T.

Organisms and the Importance of Ozonized Oils for Disinfection) (Reported at a meeting of the Society of Russian Physicians in St. Petersburg on May 10, 1884), *Trudy Obshchestva russkikh vracheĭ v S.-Peterburge s prilozheniem Protokolov zasedaniĭ Obshchestva za 1883–1884 gg. (Proceedings of the Society of Russian Physicians in St. Petersburg with a Supplement of the Proceedings of the Meetings of the Society for 1883–1884)*, St. Petersburg, Issue 2, 374–375 (1885).

41. O sostave kirpichnogo chaia (On the Composition of Brick Tea) (Reported at a meeting of the Society for Russian Physicians on May 17, 1884), *Trudy Obshchestva russkikh vracheĭ v S.-Peterburge s prilozheniem Protokolov zasedaniĭ Obshchestva za 1883–1884 gg. (Proceedings of the Society of Russian Physicians in St. Petersburg with a Supplement of the Proceedings of the Meetings of the Society for 1883–1884)*, St. Petersburg, Issue 2, 372–374 (1885); a modification of the article was published in the journal *Zdorov'e (Health)*, 1883 (see No. 38 above).

1886

42. Uproshchennyĭ azotometricheskiĭ sposob opredeleniia [mocheviny i] azota v primenenii k klinicheskomu opredeleniiu metamorfoza azotistykh veshchestv v organizme s sovremennoĭ tochki zreniia (A Simplification of an Azotometric Method for the Determination of [Urea and] Nitrogen Applied to a Clinical Determination of the Metamorphosis of Nitrogeneous Substances in an Organism from a Contemporary Point of View), [Tipografiia Ya. Trei,] St. Petersburg, 1886; also in *Voenno-meditsinskiĭ zhurnal (Military-Medical Journal)*, *155* (6), 5–38 (1886).

A List of Borodin's Musical Compositions

I. Musical-Dramatic Works

1867–1868. Choruses from and unfinished opera *Tsarskaia nevesta* (The Czar's Bride) (on a subject of L. A. Mei). [Unpublished. Original manuscripts not preserved. — T.]

1869–1887. *Kniaz' Igor'* (Prince Igor). Opera in four acts with a prolog. Libretto by A. P. Borodin (based on a reworked scenario by V. V. Stassov). Finished and orchestrated by N[ikolaĭ] A[ndreevich] Rimsky-Korsakov and A[leksandr] K[onstantinovich] Glazunov. [First performance, St. Petersburg on November 4, 1890. Published by Beliaev in 1889. Dedicated to Glinka's memory. Manuscript is in GPB[1]; several sketches are in GITIM. Polovtsian Dances (piano score) is in GTMMK. Various sketches are in LGK and in private hands. Themes from the opera were used by Robert Craig Wright and George Forrest in their musical *Kismet*, which was based on Edward Knoblock's play of the same name and which premiered at the Ziegfeld Theatre in New York City on December 3, 1953. For example, themes from the Polovtsian Dances became the songs *Stranger in Paradise* and *He's in Love*; music from the overture was used in the song *The Olive Tree*; music from Act I, Scene 1 and Konchak's aria from Act II were used in *Baubles, Bangles, and Beads* and *Gesticulate*, respectively. Wright and Forrest used a number of the songs from *Kismet* in the musical *Timbuktu*, also based on Knoblock's play. It premiered at the Mark Hellinger Theatre in New York City on March 1, 1978. — T.]

1872. Fourth (concluding) act of an opera-ballet *Mlada*, written together with C. A. Cui, M. P. Mussorgsky, and N. A. Rimsky-Korsakov. Libretto by [V. A. Krylov. Scenario by] S. A. Gedeonov. The score of the finale was published [by Beliaev] after the author's death in an orchestration by N. A. Rimsky-Korsakov. [Never produced. Original manuscripts of Nos. 2–8 are in GPB. — T.]

1866–1867. *Bogatyri* (The Valiant Knights). A musical-historical drama from prehistoric times in five scenes of everyday life by V. Aleksandrov (V. A. Krylov). Orchestration by E. N. Merten. [First produced at the Bolshoi Theatre on November 6, 1867. Unpublished manuscript is in TBAKT. A sketch of the finale to Scene 5 is found in the Library of Manuscripts at Stanford University. — T.]

II. Symphonic Works

1862–1867. Symphony No. 1 (E Flat Major).
I. Adagio [(E Flat Minor)]. Allegro [(E Flat Major)].

---

[1] Capital letters designate the present location of original manuscripts. The abbreviations are GDMC — Gosudarstvennyĭ Dom-Muzeĭ P. I. Chaikovskogo v Klinu (The State Museum at Tchaikovsky's House, Klin); GITIM — Gosudarstvennyĭ Institut Teatra i Muzyki v Leningrade (The State Institute of Theatre and Music in Leningrad); GPB — Gosudarstvennaia Publichnaia Biblioteka imeni M. E. Saltykova-Shchedrina v Leningrade (The Saltykov-Shchedrin State Public Library, Leningrad); GTMMK — Gosudarstvennyĭ Tsentral'nyĭ Muzeĭ Muzykal'noĭ Kul'tury imeni M. I. Glinki v Moskve (The Glinka State Central Museum of Musical Culture, Moscow); LGK — Leningradskaia Gosudarstvennaia Konservatoriia imeni Rimskogo-Korsakova (The Rimsky-Korsakov Leningrad State Conservatory); NIITMK — Nauchno-Issledovatel'skiĭ Institut Teatra, Muzyki i Kinematografii (Leningrad) (The Research Institute for Drama, Music, and the Cinema, Leningrad); and TBAKT — Tsentral'naia Muzykal'naia Biblioteka Akademicheskikh Teatrov (The Central Musical Library of Academic Theatres). — T.

II. Scherzo (Prestissimo) [(E Flat Major)].
[Trio (Allegro) (G Sharp Minor)].
III. Andante [(D Major)].
IV. Allegro molto vivo [(E Flat Major)].
[Dedicated to Balakirev. First performance at a Free Music School concert on December 19, 1865. Arrangement for piano duet is in GTMMK. Variant of first movement is in GPB. Sketches of score are in NIITMK. The first theme of the fourth movement was used by Wright and Forrest for the song *Gesticulate* in their musical *Kismet*. — T.]

1869–1876. Symphony No. 2 (B Minor).
I. Allegro [moderato (B Minor)].
II. Scherzo [(Molto vivo) (F Major). Trio (Allegretto) (D Major)].
III. Andante [(D Flat Major)].
IV. Allegro [(B Major)].
[Dedicated to Ekaterina Sergeevna Borodina (Borodin's wife). First performance at a Russian Musical Society concert on February 26, 1877. Published by Bessel in 1887. Fragments of the manuscript are in LGK. Complete final draft of the score is in GPB. The first theme of the first movement was used by Wright and Forrest for the song *Fate* in their musical *Kismet*. — T.]

1880. V Sredneĭ Azii (In the Steppes of Central Asia). A musical scene [Allegro con moto (A Minor — A Major). Dedicated to Franz Liszt. First performance by Orchestra of the Russian Opera conducted by Rimsky-Korsakov on April 8, 1880. Piano duet and full score are in GPB. Original manuscript is in GTMMK. Sketches for themes are in LGK. A theme was used by Wright and Forrest for the song *Sands of Time* in their musical *Kismet*. — T.]

1886–1887. Symphony No. 3 (A Minor) ("Unfinished").
I. Moderato assai [(A Minor — A Major)]. This and the next movement were finished according to an outline, orchestrated by A. K. Glazunov, [and published by Beliaev in 1888].
II. Scherzo (Vivo) [(D Major). Trio (Moderato) (B Major)].
III. Theme with variations [(Andante) (C Minor?)]. Not preserved.
IV. Finale.
[The second movement was adapted for string quartet, *Les vendredis*. Scherzo (5/8, without trio) is in GPB; sketches are in LGK and in S. A. Dianin's possession. — T.]

III. Chamber-Instrumental Ensembles

1847. Concerto for flute with piano accompaniment (D [Major — D] minor). Not preserved.

1848. [First String] Trio for two violins and 'cello on themes from *Robert le diable* (Robert the Devil) by [Giacomo] Meyerbeer. Not preserved.

1855. [Third String] Trio on the theme of a Russian song Chem tebia ya ogorchila? (How Have I Offended Thee?) for two violins and 'cello (G Minor). Published for the first time in 1946 in Borodin's Complete Works (Vol. V, issue 4), edited by P. A. Lamm. [Dedicated to P. I. Vasiliev. Original manuscript and transcription for piano duet by M. R. Shchiglev are in NIITMK. Score and parts were published by Muzgiz in 1946. — T.]

1860. Sextet (unfinished?) for two violins, two violas, and two 'cellos (D minor).
I. Allegro [D Minor].
II. Andante [E Minor].
Published for the first time in 1938 (Complete Works, Vol. V, issue 3). [Sketch for the introduction to the finale (D Major) is in GPB. Forty pages of the manuscript are in NIITMK. The score was published by Muzgiz in 1947. The last two movements are missing but must have been written. — T.]

1862. Piano Quintet for two violins, viola, and 'cello (C Minor).
I. Andante [(C Minor)].
II. Scherzo (Allegro non troppo) [(A Minor)].
III. Finale [(C Minor — C Major)].
Published for the first time in 1938 [by Iskuss'tvo] (Complete Works, Vol. V, issue 5). [Original manuscript is in GPB. — T.]

1874–1879. String Quartet No. 1 for two violins, viola, and 'cello (A major).
I. Moderato. Allegro ([A Major)].
II. Andante con moto ([F Sharp Minor). Fugato, un poco più mosso (D Minor)].
III. Scherzo (Prestissimo) [(F Major). Trio (Moderato) (A Major)].
IV. [Finale] (Andante. Allegro risoluto) [(A Minor — A Major)].
[Dedicated to Nadezhda Nikolaevna Rimskaia-Korsakova (née Purgold), Nikolaĭ Andreevich Rimsky-Korsakov's wife. The first public performance took place on December 30, 1880. Manuscript sketches are in GPB and LGK. Score and piano duet transcription were published by D. Rahter in 1884 and 1887, respectively. The second theme of the fourth movement was used by Wright and Forrest for the song *Was I Wazir?* in the musical *Kismet*. — T.]

1881–1882. String Quartet No. 2 for two violins, viola, and 'cello (D major).
I. Allegro moderato [(D Major)].
II. Scherzo (Allegro) [(F Major)].
III. Nocturne (Andante) [(A Major)].
IV. [Finale] (Andante-Vivace) [(D Major)].
[Dedicated to Ekaterina Sergeevna Borodina (née Protopopova), Borodin's wife. The first performance took place on January 26, 1882 at a chamber music concert of the Russian Musical Society. Original manuscript is in GPB, and sketches are in LGK. The score was published by Beliaev in 1888. A transcription for piano duet was made by S. M. Blumenfeld. Themes from the second and third movements were used by Wright and Forrest in the songs *Baubles, Bangles, and Beads* and *This Is My Beloved*, respectively in their musical *Kismet*. — T.]

1886. Serenata alla Spagnola (Spanish Serenade). Andante for String Quartet [in B Flat Major], written by Borodin together with Rimsky-Korsakov [(First Movement)], Liadov [(Scherzo)], and Glazunov [(Finale)] and dedicated to M[itrofan] P[etrovich] Beliaev [on the occasion of his name day, which took place on November 23, 1886. The score, parts, and piano-duet arrangement were published by Beliaev in 1887. The original manuscript is in GPB, and the sketches are in LGK. — T.]

IV. Works for Piano

1843. Polka (Hélène) [(D Minor) for piano duet]. Published for the first time [by Muzgiz] in 1946.
Youthful fugues, variations, etc. Not preserved.

1853. Scherzo (B Minor). Not preserved.

1861. Tarantella for piano for 4 hands [(D Major). Autographed copy minus last page is in GPB. Last page is in NIITMK. Published by Muzgiz in 1938. — T.]

1880. Polka, Funeral March, Requiem, and Mazurka from *Parafrazy* (Paraphrases) [for piano duet] on a comic children's theme, composed by Borodin together with Rimsky-Korsakov, Cui, Liadov, and Liszt. [Manuscript is in GPB; sketch of Polka is in NIITMK; sketch of Requiem is in LGK. Published by D. Rahter in 1879. — T.]

1885. Malen'kaia siuita (Petite Suite) [, subtitled Petit poème d'amour d'une jeune fille. — T.]
1. V monastyre (Au Couvent) [(Andante religioso) (C Sharp Minor) (The Church's vows foster thoughts only of God). — T.]
2. Intermezzo [(Tempo di menuetto) (F Major) (Dreaming of Society Life). — T.]
3. Mazurka I [(Allegro) (C Major) (Thinking Only of Dancing). — T.]
4. Mazurka II [(Allegretto) (D Flat Major) (Thinking Both of the Dance and the Dancer). — T.]
5. Grezy (Rêverie) [(Andante) (D Flat Major) (Thinking Only of the Dance). — T.]
6. Serenada (Sérénade) [(Allegretto) (D Flat Major) (Dreaming of a Love Song). — T.]
7. Nocturne [(Andantino) (G Flat Major) (Lulled by the Happiness of Being in Love). — T.]

[Dedicated to Louise, Countess of Mercy-Argenteau. Published by Bessel in 1885. The corrected proofs with Borodin's corrections are in GPB. Autograph manuscripts of Nos. 2–6 are in the British Museum. — T.]
Scherzo (A Flat Major). The first publication indicates that it is the author's orchestral arrangement. However, neither score nor orchestral outline are known to us. [Dedicated to the Belgian composer

Théodore Jadoul. Glazunov orchestrated it as No. 7 (Finale) for the Petite Suite. The Serenade was used by Wright and Forrest for the song *Night of My Nights* in their musical *Kismet*. — T.]

V. Vocal Works

Razliubila krasna devitsa (The Fair Maiden No Longer Loves Me) [(with 'cello obbligato in D Major). Manuscript is in NIITMK. Published by Muzgiz in 1947. — T.]
Slushaĭte, podruzhen'ki, pesenku moiu (Friends, Hear My Song) [(voice and piano with 'cello obbligato in E Minor). Manuscript is in NIITMK. Published by Muzgiz in 1947. — T.]
Chto ty rano, zoren'ka? (Why Art Thou So Early, Dawn?) [Arrangement of folk song (F Sharp Minor). Manuscript is in GPB. Published by Muzgiz in 1947. — T.]

1855. Krasavitsa rybachka (The Beautiful Fishermaiden) (words by H[einrich] Heine) [D Flat Major). Dedicated to A. S. Shashina. Also arrangement with 'cello obbligato. Manuscript is in NIITMK. Published by Muzgiz in 1947. — T.]

1867. Spiashchaia kniazhna (The Sleeping Princess) (Words by A. Borodin) [A Fairy tale for voice and piano (A Flat Major). Dedicated to N. A. Rimsky-Korsakov. Manuscripts are in GPB and GTMMK. Published by Jurgenson in 1870. Score and parts (orchestrated by Rimsky-Korsakov) published by Jurgenson in 1904. — T.]

1868. Morskaia tsarevna (The Sea Princess) (words by A. Borodin) [Ballad for voice and piano (F Major). Dedicated to A. E. Makovskaia. Manuscripts are in GPB and GTMMK. Published by Bessel in 1873. — T.]

1868. Pesnia temnogo lesa (Song of the Dark Forest) (words by A. Borodin) [(F Sharp Minor). Dedicated to L. I. Shestakova, Glinka's sister. Manuscripts are in GPB and GTMMK. Published by Bessel in 1873. Score and parts, edited by Glazunov, published by Bessel in 1893 for male chorus with piano. — T.]

1868. Fal'shivaia nota (The False Note) (words by A. Borodin) [(D Flat Major). Dedicated to M. P. Mussorgsky. Manuscripts are in GPB and GTMMK. Published by Jurgenson in 1870. — T.]

1868. Otravoĭ polny moi pesni (My Songs Are Filled With Poison) (words by H[einrich] Heine) [Ballad for tenor and piano (E Flat Minor). Dedicated to C. A. Cui. Manuscript is in GTMMK. Published by Jurgenson in 1870. — T.]

1870. More (The Sea) (words by A. Borodin) [Ballad for tenor and piano (G Sharp Minor). Dedicated to V. V. Stassov. Manuscripts are in GPB and GTMMK. Published by Jurgenson in 1870. Orchestrated by N. A. Rimsky-Korsakov and published by Jurgenson in 1904. — T.]

1873. Iz slez moikh (From My Tears) (words by H[einrich] Heine). [Ballad for voice and piano (B Major). Dedicated to M. S. Stupishina. Manuscript is in GDMC. Published by Bessel in 1873. — T.]

1881. Dlia beregov otchizny dal'noĭ (For the Shores of a Distant Homeland) (words by A[leksandr Sergeevich] Pushkin). [Ballad for voice and piano (C Sharp Minor). Dedicated to Ekaterina Sergeevna Borodina (née Protopopova), Borodin's wife. Manuscript is in GPB. Published by Beliaev in 1890. — T.]

1882. Arabskaia melodiia (Arab Melody) (words by A. Borodin). [Ballad for voice and piano (F Major). Published by Beliaev in 1889. — T.]

1884. U liudeĭ-to v domu (At Home Among Real People) (words by N[ikolaĭ Alekseevich] Nekrasov) [Song with piano or orchestral accompaniment (F Major). Dedicated to Daria M. Leonova. The orchestral score is in GPB; rough sketches and fragments are in LGK. The score and parts and an edition with piano accompaniment by G. O. Dutsch were published by Beliaev in 1890. — T.]

1884. Spes (Pride) (words by A. K. Tolstoy). [Ballad for voice and piano (F Major). Dedicated to Anna A. Bichurina. The manuscript is in GPB; a copy of the autograph with Borodin's comments is in LGK. Published by Beliaev in 1890. — T.]

1885. Chudnyĭ sad (The Wonderful Garden). Septain (heptameter) by the Belgian poet G. C. [(Georges Collen) (in French). Ballad for voice and piano (D Flat Major). Dedicated to Louise, Countess of

Mercy-Argenteau. Rough sketches are in LGK. Published by Veuve Muraille, Liège in 1885. Published in Russian translation by Bessel in 1887. — T.]

Slava Kirillu! (Glory to Cyril!). For male chorus [with piano accompaniment (unfinished, unpublished). — T.]

Serenada chetyrekh kavalerov odnoĭ dame (Serenade in Honor of One Lady by Four Cavaliers). Vocal quartet (words by A. Borodin) [for four men's voices with piano accompaniment (D Flat Major). — T.]

Principal Literature on Borodin

1. Pis'ma A. P. Borodina (Letters of A. P. Borodin), [edited by S. A. Dianin,] Volume I (1857–1871), [Gosudarstvennoe izdatel'stvo, muzykal'nyĭ sektor,] Moscow, 1927–1928.
2. Pis'ma A. P. Borodina (Letters of A. P. Borodin), [edited by S. A. Dianin,] Volume II (1872–1877), [Gosudarstvennoe muzykal'noe izdatel'stvo,] Moscow, 1936.
3. Pis'ma A. P. Borodina (Letters of A. P. Borodin), [edited by S. A. Dianin,] Volume III (1878–1882), [Gosudarstvennoe muzykal'noe izdatel'stvo,] Moscow, 1949.
4. Pis'ma A. P. Borodina (Letters of A. P. Borodin), [edited by S. A. Dianin,] Volume IV (1883–1887), [Gosudarstvennoe muzykal'noe izdatel'stvo,] Moscow, 1950.
5. A. P. Dianin: A. P. Borodin. Biograficheskiĭ ocherk i vospominaniia (A. P. Borodin. Biographical Sketch and Reminiscences), *Zhurnal Russkogo fiziko-khimicheskogo obshchestva (Journal of the Russian Physico-Chemical Soviety)*, *20* (4), 367–379 (1888).
6. V. V. Stassov: A. P. Borodin. Ego zhizn', perepiska i muzykal'nye stat'i (1834–1887) (A. P. Borodin. His Life, Correspondence, and Musical Articles (1834–1887)), [Souvorin,] St. Petersburg, 1889 [See Item No. 23, Additional Literature. — T.]
7. V. V. Stassov: Biografiia A. P. Borodina (Biography of A. P. Borodin), *Istoricheskiĭ vestnik (Historical Herald)*, St. Petersburg, Vol. 28 (1887).
8. V. V. Stassov: 25 let russkogo iskusstva (25 Years of Russian Art), *Vestnik Evropy (European Herald)*, [October,] 1883 [For an English translation see Vladimir Vasilevich Stasov: Selected Essays on Music, translated by Florence Jonas; introduced by Gerald Abraham, Barrie & Rockliff, The Cresset Press, London, 1968, pp. 66–116. — T.]
9. M[ikhail] Nikolaevich Mladenstsev and V. E. Tishchenko: D[mitriĭ] I[vanovich] Mendeleev, [ego zhizn' i deiatel'nost [Dmitriĭ Ivanovich Mendeleev, His Life and Work)], [Akademiia Nauk SSSR, Moscow,] Vol. I, 1938.
10. Istoriia Imperatorskoĭ voenno-meditsinskoĭ (byvsheĭ Mediko-khirurgicheskoĭ) akademii za sto let (1798–1898) (History of the Imperial Military-Medical (formerly Medical-Surgical) Academy during the Hundred Years (1798–1898)), St. Petersburg, 1898.
11. E[lena] I[osipovna] Likhacheva: Materialy dlia istoriĭ zhenskogo obrazovaniia v Rossi (1856–1880) (Materials for the History of Higher Education for Women in Russia (1856–1880)), [Tipografiia M. M. Stasiulevicha,] St. Petersburg, [1899–]1901.
12. O khimicheskikh rabotakh A. P. Borodina (On the Chemical Works of A. P. Borodin), *Zhurnal Russkogo khimicheskogo obshchestva (Journal of the Russian Chemical Society)*, *1*, 11, 31, 214, 225 (1869); *2*, 91, 254 (1870); *3*, 7, 64, 127 (1871); *4*, 60, 207–209 (1872); *5* (8), 383–389, 409 (1873); *6*, 10, 94, 176, 251 (1874); *7*, 9–10, 282, 285–288 (1875); *8*, 9, 43, 145, 280 (1876); *9* (6), 240 (1877); *10*, 324 (1878); *11*, 86 (1879).
13. N. A. Rimsky-Korsakov: Letopis' moeĭ muzykal'noĭ zhizni (Chronicle of My Musical Life), 5th Edition, [Gosudarstvennoe muzykal'noe izdatel'stvo,] Moscow, 1935.
14. T. V. Volkova: Pis'ma Borodina k D. I. Mendeleevu (Letters of Borodin to D. I. Mendeleev), *Uspekhi khimii (Advances in Chemistry)*, *9* (9), 1060–1071 (1940).
15. Yu. I. Solov'ev: Raboty A. P. Borodina po al'dol'noĭ kondensatsii (A. P. Borodin's Works on the Aldol Condensation), *Uspekhi khimii (Advances in Chemistry)*, *18* (6), 756–759 (1949).

Additional Literature on Borodin

Compiled by George B. Kauffman

1. Anonymous: Obituary: Professor Borodin, *The Lancet*, *1*, 601 (March 19, 1887).
2. B. W. Feddersen and A. J. Oettingen (eds.): Poggendorff's Biographisch-Literarisches Handwörterbuch, Johann Ambrosius Barth, Leipzig, 1898, Vol. 3, pp. 164–165; 1904, Vol. 4, p. 160.
3. Paul Rosenfeld (ed.): Musical Portraits: Interpretations of Twenty Modern Composers, Harcourt, Brace & Howe, New York, 1920, pp. 149–158.
4. Louise Cruppi: Borodin and Liszt, *Living Age*, *312*, 600–605 (March 11, 1922).
5. Gerald E. H. Abraham: Borodin: The Composer & His Music: A Descriptive and Critical Analysis of His Works and a Study of His Value as an Art-Force, William Reeves, Bookseller Ltd., London, [1927].
6. Alexander [Ivanovich] Petrunkevitch: Russia's Contribution to Science, *Transactions of the Connecticut Academy of Arts and Sciences*, *23*, 223 (June, 1930).
7. Frederick H. Getman: Alexander Borodin — Chemist and Musician, *Journal of Chemical Education*, *8* (9), 1762–1780 (September, 1931).
8. Hermann Bernhard: Alexander Borodin, *Journal of Chemical Education*, *9* (1), 153–154 (January, 1932).
9. Gerald Abraham: Borodin's Songs, *Musical Times*, *75*, 983–985 (1934).
10. Anonymous: Chemistry and Music, *The Etude*, *52*, 508 (September, 1934).
11. Anonymous: A. P. Borodin (1834–87), *Nature*, *134*, 727 (November 10, 1934).
12. Hope Stoddard: Borodin, Genius in Double-Harness, *Musical Opinion*, *57*, 502–503 (1934).
13. M. M. Kurbanoff (translated and arranged by Alfred J. Swan): A Few Reminiscences of Borodin (1884–87), *The Chesterian*, *16*, 96–99 (1935).
14. Gerald Abraham: Studies in Russian Music, William Reeves, London, [1936], Chap. 6, Borodin as a Symphonist (pp. 102–118), Chap. 7, Prince Igor (pp. 119–141).
15. R. P. La Combe: Le cas de Borodine, Thesis, Maloine, Paris, 1936.
16. Michel D. Calvocoressi: In: Michel D. Calvocoressi and Gerald E. H. Abraham: Masters of Russian Music, Duckworth, London, 1936, Alfred A. Knopf, New York, 1936, pp. 155–177.
17. F. William Sunderman: Alexander Porfirivich Borodin: Physician, Chemist and Composer, *Annals of Medical History*, *10*, 445–453 (1938).
18. George Sarton: Borodin (1833–87), *Osiris*, *7*, 225–260 (1939).
19. Harold B. Friedman, "Alexander Borodin — Musician and Chemist," *Journal of Chemical Education*, *18* (11), 521–525 (November, 1941).
20. Edmund Yochum: Borodin: Symphonies and Syntheses, *The Science Counselor*, *7*, 42–43, 59–60 (1942).
21. Victor I. Seroff: The Mighty Five: The Cradle of Russian National Music, Allen, Towne & Heath, New York, 1948.
22. Mikhail Osipovich Zetlin: Piatero i drugie, 2nd edition, Izdatel'stvo M. E. Tsetlin, New York, 1953; translated into English and edited by George Panin as: The Five: The Evolution of the Russian School of Music, International Universities Press, New York, 1959; reprinted by Greenwood Press, Westport, Connecticut, 1975.
23. Vladimir Vasil'evich Stassov: Aleksandr Porfir'evich Borodin, Gosudarstvennoe muzykal'noe izdatel'stvo, Moscow, 1954. An earlier printing of this book was translated into French and then into English: Alfred Habets: Alexandre Borodine, d'après la biographie et la correspondence publiées par Wladimir Stassoff, Fischbacher, Paris, 1893; translated into English with a preface by Rosa Newmarch as: Borodin and Liszt: I. Life and Works of a Russian Composer, II. Liszt, as Sketched in the Letters of Borodin, 2nd edition, Digby, Long & Co., London [,1895]; reprinted by AMS Press, New York, 1977.

24. Jerzy Chodkowski: Aleksander Borodin jako chemik (Alexander Borodin as a Chemist), *Wiadomości Chemiczne*, *8*, 369–373 (1954) (in Polish).
25. John Briggs: The Collector's Tchaikovsky and the Five, J. B. Lippincott Co., Philadelphia and New York, 1959, pp. 164–181.
26. Sergeĭ Aleksandrovich Dianin: Borodin, zhizne opisanie, materialy i dokumenty, 2nd edition, Gosudarstvennoe muzykal'noe izdatel'stvo, 1960, translated into English by Robert Lord as: Borodin, Oxford University Press, London, New York, Toronto, 1963.
27. David Lloyd-Jones: Borodin in Heidelberg, *The Musical Quarterly*, *46* (4), 500–508 (1960).
28. David Lloyd-Jones: Borodin on Liszt, *Music and Letters*, *42* (2), 117–126 (1961).
29. M. Schofield: Pioneers in Aldehydes and Ketones, *Perfumery and Essential Oil Record*, *56* (7), 451–452 (July, 1965).
30. Walter Kwasnik: Der Komponist Alexander Borodin (1834–1887) als Chemiker: Zur Wiederkehr seines Todestages am 28. 2. 1967, *Chemiker-Zeitung/Chemische Apparatur*, *91* (9), 312–313 (1967).
31. Ward Botsford: Angel's 'Prince Igor', *American Record Guide*, *34*, 358–361 (January, 1968).
32. P. De Ceuster: Alexander Borodin, chemisch musicus of muzikal chemicus (Alexander Borodin, Chemical Musician or Musical Chemist), *Chemisch Weekblad*, *65* (10), 11–12 (1969) (in Dutch).
33. Jean-Albert Gautier: Comment les découvertes du chimiste Kekulé empêchèrent Borodine de terminer le 'Prince Igor', *Revue d'Histoire de la Pharmacie*, *20* (No. 204), 5–10 (March, 1970).
34. V. A. Kiselov: Novye pis'ma Borodina (Some New Letters of Borodin's). In: Mikhail Pavlovich Alekseev et al. (eds.), Muzykal'noe nasledstvo (Musical Heritage), Moscow, 1970, Vol. 3, pp. 208–239.
35. V. A. Kiselov: Stsenicheskaia istoriia pervoĭ postanovki Kniazia Igoria (The History of the First Performance of Prince Igor). In: Mikhail Pavlovich Alekseev et al. (eds.): Muzykal'noe nasledstvo (Musical Heritage), Moscow, 1970, Vol. 3, pp. 284–352.
36. Y. I. Soloviev: Aleksandr Porfirevich Borodin. In: Charles Coulston Gillispie (ed.): Dictionary of Scientific Biography, Charles Scribner's Sons, New York, 1970, Vol. 2, pp. 316–317.
37. Charlene Steinberg: The Scientific Activities of Aleksandr Borodin, *CHEM TECH*, *1*, 473–475 (August, 1971).
38. E. Lee Strohl, Robert W. Jamieson, and W. G. Diffenbaugh: Physicians — Musicians, *Perspectives in Biology and Medicine*, *17* (2), 267–285 (Winter, 1974).
39. Gerald Abraham: Arab Melodies in Rimsky-Korsakov and Borodin, *Music and Letters*, *56* (3–4), 313–318 (1975).
40. Maurice Schofield: Borodin — Chemist and Composer, *Chemistry*, *49* (8), 13–14 (October, 1976).
41. Gerald Abraham and David Lloyd-Jones: Alexander Porfir'yevich Borodin. In: Stanley Sadie (ed.): The New Grove Dictionary of Music and Musicians, Macmillan Publishers Ltd., London, 1980, Vol. 3, pp. 54–62.
42. Martin Sherwood: A Russian of Many Octaves, *New Scientist*, *100*, 424 (November 10, 1983).
43. Gerald Seaman: Borodin's Letters, *The Musical Quarterly*, *70* (4), 476–498 (Winter, 1984).
44. George B. Kauffman: Russia's Aleksandr Borodin: Many gifts, many callings, *Industrial Chemist*, *8* (1), 40–43 (January, 1987).
45. George B. Kauffman, Ian D. Rae, Yuriĭ Ivanovich Solov'ev, and Charlene Steinberg: Borodin: Composer and Chemist, *Chemical and Engineering News*, *65* (7), 28–35 (February 16, 1987).
46. George B. Kauffman: Though not prolific, Borodin was multitalented, *The Fresno Bee*, February 22, 1987, p. F 4.
47. Alvan D. White: Alexander Borodin: Full-Time Chemist, Part-Time Musician, *Journal of Chemical Education*, *64* (4), 326–327 (April, 1987).
48. Clive B. Hunt: Aleksandr Borodin: chemist and composer, *Chemistry in Britain*, 23 (6), 547–550 (June, 1987).
49. George B. Kauffman, Yuriĭ Ivanovich Solov'ev, and Charlene Steinberg: Aleksandr Porfir'evich Borodin (1834–1887), *Education in Chemistry*, 24 (5), 138–140 (September, 1987).
50. N. A. Figurovskiĭ and Yu. I. Solov'ev: Aleksandr Porfir'evich Borodin, Izdatel'stvo Akademii Nauk SSSR, Moscow, Leningrad, 1950; Aleksandr Porfir'evich Borodin: A Chemist's Biography, translated into English by Charlene Steinberg and George B. Kauffman, Springer-Verlag, Berlin, Heidelberg, New York, 1988.

51. George B. Kauffman and Kathryn Bumpass: The Apparent Conflict between Art and Science: The Case of Aleksandr Porfir'evich Borodin (1833–1887), *Leonardo*, in press.
52. George B. Kauffman: Syntheses and Symphonies, *The World & I*, 3(1), 206–211 (January, 1988).
53. Kathryn Bumpass and George B. Kauffman: Nationalism and Realism in 19-th Century Russian Music: 'The Five' and Borodin's Operatic Masterpiece, *Prince Igor*, *The Music Review*, in press.
54. Ian D. Rae: Borodin's Chemical Research: A Retrospective View on the Centenary of His Death, *Ambix*, in press.

Appendices

Appendix I

Excerpt from the Report of the Meeting of the Conference of the St. Petersburg Medical-Surgical Academy, September 19, 1859[1]

Having completed the course in medical sciences as physician with distinction in 1856, Aleksandr Borodin attracted attention throughout the entire course both by his distinctive abilities and by his love for science. After the doctor's examination he was retained for service at the 2nd Military-Land Forces Hospital with attachment to the Department of General Therapy, Pathology, and Clinical Diagnostics. In 1858 he passed the examination for the degree of Doctor of Medicine, and since he showed a special inclination toward the pursuit of the natural sciences, especially chemistry, he was retained as assistant in the Department of Chemistry to direct the practical work of the students of the 2nd course and of young medical doctors who remained for advanced courses in science at the 2nd Land Forces Hospital. In addition to the duties assigned to him, Mr. Borodin carried out a chemical investigation of the mineral waters in the town of Soligalich in Kostroma Province and, carrying out chemical research, presented two articles to the Academy of Sciences: (1) Recherches sur la constitution chimique de l'hydrobenzamide et de l'amarine [*B 1*]; (2). Über die Wirkung des Jodaethyls auf Benzoylanilid [*B 3*], which were approved and published in the Bulletin de l'Academie de St. Pétersbourg.

In addition to this, according to regulations imperially approved on May 21 and followed by imperial command on June 10, 1858, Borodin was appointed to be in the number of those retained at the 2nd Miliary-Land Forces Hospital for further training for a 3-year period.

The Conference of the Academy, being convinced of Doctor Borodin's unquestionable talents by both his teaching and his scientific works and bearing in mind that his 3-year period at the hospital is expiring, unanimously decided to petition the President of the Academy to send him abroad for 2 years with payment, in addition to the allowance received for service at the 2nd Military-Land Forces Hospital, of 1000 silver rubles from the sums set aside for trips abroad of young doctors, retained for improvement at the hospital and the Academy.

Footnote to Appendix I

1 Military-Historical Archives of the USSR in Leningrad, Stock No. 749, Military-Medical Academy, File No. 194 for 1859, Sheets 190–191. — A.

Appendix II

Zinin's Letter of Instructions to Doctor Borodin[1]

The Conference of the Imperial Medical-Surgical Academy, in sending you abroad for improvement in chemistry, imposes upon you the duty to carry out the following:

1. During your stay abroad to study thoroughly several special research methods which are particularly important both for pure and for applied chemistry, i.e., methods of gas analysis,

technique of analysis by titration of solutions, carrying out of chemical reactions in sealed tubes under high pressure, and carrying out of operations at high temperatures; for this goal you must visit the laboratories of Bunsen in Heidelberg, Wurtz, Berthelot, and Sainte-Claire Deville in Paris, and Hofmann in London.
2. To bear constantly in mind the application of chemistry to physiological and medical sciences, and for this goal you must visit the laboratories of Scherer in Würzburg and Liebig in Munich.
3. To inspect European factories and plants, which are most remarkable from a chemical point of view, i.e., chemical, gas, sugar, glass, etc., and plants for the smelting of metals in England, France, Germany, and Belgium.
4. To look at major localities of Europe which are important in chemical, mineralogical, and geognostical[2] respects, i.e., sites of principal mineral ores, borax, and sulfur in Silesia, Bohemia, Hungary, Germany, France, and Italy.
5. To present an account of your work every 6 months.

(written to Dr. Borodin on November 13, 1859) [Professor Zinin]

Footnotes to Appendix II

[1] Military-Historical Archives of the USSR in Leningrad, Stock No. 749, Military-Medical Academy, File No. 12 for 1858–1863, Sheet 21. — A.

[2] Geognosy is the branch of geology that deals with the constituent parts of the earth, its envelope of air and water, its crust, and the conditions of its interior. — T.

Appendix III

Borodin's Letter of Account to the Head of the Medical-Surgical Academy, P. A. Dubovitskii[1]

Your permanent arrangement with me allows me to hope that it will be agreeable to your Excellency to know about my activities abroad. Acting on your flattering permission to write directly to your Excellency about everything which concerns me, I consider it my first duty to report my work in Heidelberg to you.

Constantly bearing in mind my main assignment – that of becoming a teacher of chemistry, I, wothout doubt, first of all had to look after the reinforcement of my knowledge in the field of this science. Therefore I first made up my mind to abandon all applied sciences and to study chemistry exclusively, and chiefly the practical study of those chemical methods with which I did not have the opportunity of becoming acquainted in [St.] Petersburg.

I intended to begin with the study of Bunsen's gasometric methods,[2] and with this goal in mind I decided to work in his laboratory. This proved to be inconvenient. Bunsen himself is now occupied with the application of a gas flame to the qualitative and quantitative determination of potassium, sodium, lithium, etc. in various minerals.[3] These studies are not of any interest to me, and therefore working with Bunsen would be of no value to me. Only a beginner can work successfully in the public laboratory of the university because this laboratory is intended primarily for the study of analytical chemistry. Furthermore, working here is rather inconvenient. The number of workers is so great that much time is wasted – in waiting one's turn to use the ovens, apparatus, etc.

The equipment in a public laboratory for the most part is not good, and clearly, studying anything seriously requires having one's own equipment – from vessels to sensitive weights – otherwise the loss of time is extraordinarily large, and furthermore, no one could be convinced of the precision of the results. Finally, in the public laboratory one can work only until 5:00 P. M., and on Saturday and Sunday no work at all occurs. Since time – particularly now – is so precious to me, I decided to work in another laboratory, that of Doctor Erlenmeyer, Privat-Dozent[4] at the university. I pay double here compared to the others, but in return I have a separate room where I can work absolutely independently when and as much as I please. Nevertheless, here too I have to acquire not only materials but also many instruments and even vessels to make my work successful. Since in Heidelberg it is either impossible to obtain very many things or they are no good and expensive, I was compelled to go to Merck[5] in Darmstadt for materials and to Paris for apparatus. I stayed in Paris for 9 days because I had to order apparatus for work in sealed tubes and to learn the many details concerning this apparatus from [Marcellin] Berthelot. Berthelot was so obliging that he supplied me with his forms and models which were necessary for ordering the apparatus. In addition, in the course of my stay in Paris he showed me in every detail everything that he had in this respect that was new. I am extremely satisfied by this circumstance because it saves me the necessity of working in his laboratory next year, which I had originally planned. My first acquisition, mainly materials, was more expensive than I expected in spite of the relatively low price of everything abroad. However, I hope that the low cost of living in Heidelberg makes up for these unforeseen expenses, especially since, for the present, I do not intend to acquire any books or journals. I have the new journals at the museum for two gulden a month, and I can get all the books and journals from previous years from the university library. I am occupied here almost solely in the laboratory and mainly by those things which I did not have the opportunity to study in Russia. I have completely abandoned my work which I started in [St.] Petersburg for lack of time. I do not attend any lectures; Bunsen and Kirchhoff read too elementarily, and Helmholtz[6] reads, according to duty, a very elementary course on the history of development instead of on his physiological research.

I intend to remain in Heidelberg until May and perhaps longer if I find it useful. For the present I am not making any assumptions concerning the summer semester. I will undertake whatever appears more convenient and advantageous.

With sincere respect and deepest devotion I have the honor to be your Excellency's obedient servant.

Heidelberg, February 10, 1860 [A. Borodin]

Footnotes to Appendix III

1 Military-Historical Archives of the USSR in Leningrad, Stock No. 749, Military-Medical Academy, File No. 12 for 1858–1863, Sheets 37–38. — A.

2 See R. W. Bunsen: Gasometrischen Methoden, Friedrich Vieweg und Sohn, Braunschweig, 1857; Gasometry, Comprising the Leading Properties of Gases, translated by Henry E. Roscoe, Walton & Maberly, London 1857. — T.

3 Together with the physicist Gustav Robert Kirchhoff (1824–1887), Bunsen laid the foundation of spectral analysis in 1860 (See G. Kirchhoff and R. Bunsen: Chemische Analyse durch Spectralbeobachtungen, edited by Wilhelm Ostwald, Ostwalds Klassiker der exakten Wissenschaften No. 72, Wilhelm Engelmann, Leipzig, 1895). Their systematic application of this method led to the discovery of two new alkali metals, cesium (1860) and rubidium (1861). — T.

4 An unsalaried lecturer whose sole income is derived from fees paid by the students who enroll in his courses. — T.

5 The E. Merck chemical pharmaceutical firm in Darmstadt was formed in 1827 from the firm Engel-Apotheke, which had been founded in 1654 and had been in the possession of the Merck family since 1668. The American firm Merck & Co., Inc., Rahway, New Jersey was formed in 1927 from a branch establishment founded in 1894 by Georg(e) Wilhelm Merck. — T.

6 Hermann Ludwig Ferdinand von Helmholtz (1821–1894) was Professor of Physiology at the Universities of Königsberg, Bonn, and Heidelberg. From 1871 until his death he was Professor of Physics at the University of Berlin. For further details see R. Steven Turner: Hermann von Helmholtz. In: DSB, 1972, Vol. 6, pp. 241–253. — T.

Appendix IV

Mendeleev's Letter to A. A. Voskresenskiĭ[1]

The Chemical congress, which just ended in Karlsruhe, constituted so remarkable an occurrence in the history of our science that I consider it my duty to describe, although briefly, the sessions of the congress and the results that it accomplished.

The desire to understand and, if possible, come to an agreement over the basic contradictions existing between followers of different chemical schools served as the important cause for convening an international chemical congress. At the beginning Mr. Kekulé proposed many problems for solution: the problem of the distinction between molecule, atom, and equivalent; the problem of the magnitudes of atomic weights, i.e., to adopt Gerhardt's weight or Berzelius' weight, which was later changed by Liebig[2] and Poggendorff[3] and now accepted by the majority – next, the problem of formulas and, finally, of those forces which, according to the contemporary state of science, are considered the cause of chemical phenomena. But in the first session, occurring on September 3 (new style), the group found it impossible to elucidate such a large number of problems in such a short time and therefore decided to dwell upon only the first two. This session was opened by an introductory speech by Mr. Weltzien, professor at the well-known Technische Hochschule in Karlsruhe. Bunsen was elected president, but he declined, saying that poor hearing would prevent him from carrying out this responsibility in the proper manner. Then the host Weltzien was elected president for that day and for the following day, – Boussingault.[4] Secretaries were also appointed with one each from the four nations having the greatest number of representatives: from the French — Wurtz; from the Germans — Strecker;[5] from the English — Roscoe;[6] and from the Russians — Shishkov.[7] Kekulé was elected the fifth secretary since, according to the general opinion, he was one of the first who had the idea of holding a congress.

Kekulé stated the essence of the problems constituting the subject of the contradictions. After long debates, the group decided to form a committee, of about 30 members, in order to determine in what form to offer the problems for voting in the congress. The committee, in which Zinin, Shishkov, and I were from the Russians, met immediately upon the closing of the first session. The committee soon came to the conclusion that the entire essence of the contradictions is concentrated in the distinction between the idea of the molecule and the idea of the atom. As soon as this distinction is recognized, then immediately the doubling of formulas, which constitutes the subject of discord in the practice of science, is admitted. So, for example, hydrogen, water, [and] sulfuric acid are usually represented as H, HO [and HSO^4 — T.], when atoms of them are understood; the new school writes them as H^2, H^2O^2, [and] $H^2S^2O^8$, when molecules of them are understood.[8] Therefore it was unanimously decided to offer the first question for voting in the congress as follows: does the majority wish to accept the distinction between atom and molecule?

The possibility of reaching any completely definite ideas during the discussion concerning equivalents had to be abandoned completely. Some took for equivalents the quantities of substances replacing each other without a change of basic properties; others considered our weights as equivalents, that is, the weight ratios of chemically bonding substances; finally, a third group found that a consistent construction of the idea of equivalents is completely impossible, that it unavoidably leads to contradictions. Contradictions are complicated even more by the question of molecules. Some wished to recognize only chemical characteristics, i.e., reactions, for the definition of a molecule of each substance; others considered only physical characteristics necessary, and finally, a third group accepted the identity of both principles, i.e., they recognized both courses and found that they lead to identical results. During these discussions the most definite and, without doubt, the most original and integral opinion was expressed by a Professor Cannizzaro from Genoa. I will try to give an account of him in a few words, preserving only the essence. Of course, I am not able to transmit to you that enthusiasm, that sensible energy forming the convictions which so powerfully affected his listeners.

Cannizzaro said that no one has ever laid out the theory of equivalents across all of chemistry, not in the sense of weights but of amounts of substances replacing one another, for example HSO^4, HNO^6,[9] etc. This theory always served only as an explanation of the composition of chemical compounds in this or that particular case. It was necessary to renounce its general application as soon as the first polybasic acid, phosphoric acid, was discovered. As soon as we recognized polybasicity, we had to relinquish the desire to express equivalent amounts by formulas. Actually, NHO^6 and $P^{1/3}HO^{8/9}$ and C^2HO^4 react with the same amount of KHO^2.[10] These formulas express equivalents of the acids but, to say nothing of the fact that it is impossible to make use of them, they also cannot be accepted because an equivalent of one and the same substance is changed. When we give caustic alkali, lime, lead hydroxide, and such [substances] a formula similar to the formula of potassium hydroxide or water (H^2O^2, KHO^2), we are trying to convey their equivalents by these formulas. This is erroneous; the formulas do not have to express that meaning ascribed to them by all chemists, beginning with Berzelius. By means of formulas they always wanted to convey the amount of a substance entering into a reaction, calling this amount sometimes an atom, simple or complex, and sometimes a molecule. Considering those means by which they reached the idea of molecular weights, Mr. Cannizzaro quite correctly recognizes only one means, relative, of course. It was shown by Ampère and Avogadro on the basis of the works of Gay-Lussac, Dumas, and Mitscherlich, and recently Gerhardt and Laurent supported it; it gave new life to the chemistry of our time. This means is the determination of the volume of a substance in the gaseous or vapor state, i.e., the determination of the specific gravity of the substance in such a state. Without this means, until now we were unable to be certain about the origin of ethers because their existence does not yet demonstrate the necessity for doubling the formula of the simple alcohol, just as the existence of some double salts does not compel one to double the formula of the salts taken separately. The rule of volumes, by virtue of its simplicity and by virtue of the relationship which it establishes between the related sciences, chemistry and physics, deserves preference over other means for determining relative molecular weights. This rule can be expressed also in this way: the distance between the centers of molecules in vapors and gases is the same for all substances and depends only on pressure and temperature. The contemporary theory of heat in relation to gases rests on this very principle. Accepting it, we do not retreat from the chemical path because everything that we know until now points to agreement between molecular weights, determined according to the volume of the vapor,[11] and the amount of a substance entering into a reaction. Accepting the idea of molecules, it is easy to obtain a completely correct idea of weights (or atoms, as everyone here has called them). Actually, knowing the specific gravity of a substance in its gaseous state, we know the weight of equal volumes of the substance . . .

In all fairness, Cannizzaro's animated speech was met with general approval. Wurtz, from his side, noted that many double decomposition reactions lead to formulas identical with the formulas which follow from Cannizzaro's weights, identical with Berzelius' old weights.

On the next day the committee met again in order to decide finally in what form to propose the question for voting to the congress. It was decided to propose the question in the following form: does the congress desire and does it consider it useful to establish a distinction between the idea of a molecule, as the amount of material entering into a reaction and with corresponding physical properties, and the idea of an atom, as the least amount of a substance entering into molecules?

Kopp[12] was the president both at this and at the previous session of the committee.

The second session of the congress on September 4 was opened by an introductory speech by Boussingault; the importance of the proposed question was clearly pointed out in it. One part of the speech, where it was stated that "the question is not about a new or an old science, that science is not growing old, we are growing old," provoked general applause. Then Kekulé talked for a long time, developing elementary ideas about atoms, molecules, and equivalents. Kekulé is inclined to recognize a difference between physical and chemical molecules; he thinks that molecules of a substance acting in thermal phenomena are not those which we find in solid, liquid, and gaseous substances and are not those which enter into reactions. However, he thinks that one can obtain the precise determination of molecular weights by chemico-physical investigations. Cannizzaro spoke after him. Wurtz and Kopp noted that the question is not by what means we can determine molecular weights: the nature of the question proposed to the congress is the question of the necessity of considering a distinction in those ideas which we associate with the words – *molecule and atom*. Then there were some brief disputes not having any significant importance. Finally, the secretaries read in English, French, and German the questions proposed for voting. Here they are:

"It is proposed to accept a difference in the idea of molecule and atom, regarding the amount of a substance entering into reactions and with identifying physical properties as the molecule and regarding the least amount of a substance contained in molecules as the atom."

"Further – it is proposed to regard the equivalent as empirical and independent of the idea of atoms and molecules."

According to the president's suggestion, those who agreed to accept these positions were to raise their hands. It appeared that a majority of the group were in agreement. Then it was suggested that those who reject the proposals raise their hands. Only one hand was raised, and it was immediately lowered. The result was unexpectedly unanimous and significant. Accepting a distinction between atom and molecule, chemists of all countries accepted the principle of the unitary system. Now it would be a great inconsistency, having accepted the principle, not to accept its consequences.

Dumas arrived near the end of this session, having just come from Paris. His appearance was met with general applause. At first, Kopp was elected president of the September 5 session, but he declined. Then Dumas was unanimously elected, and he consented.

During the evening the committee met under Dumas' chairmanship. It was necessary to consider in what form to present the question about formulas and weights to the congress. Dumas, in a speech not devoid in parts of real eloquence, pointed out the complete necessity of the uniform designation of weights (i.e., to agree, for example, that C denotes either 6 or 12). He said that this is especially important for beginners and for teachers. Students, listening to different teachers, do not understand one another, confusing marsh gas with olefiant gas[13] [and] calomel with corrosive sublimate.[14] In Dumas' opinion it is impossible to devise a weight satisfying physical properties. The best method for complying with all the requirements is to take Berzelius' old weights, introducing modifications required by the new advancements. Wurtz and Cannizzaro remarked to Dumas that by introducing modifications in Berzelius' weights we obtain precisely the weights required by Gerhardt's new theory. Erdmann[15] requested complete freedom both in the designations and also in everything else and wished only that they agree on the symbols, for example, which C to take for 12 and which for 6. By general agreement it was decided to propose the question in the following form: does the congress wish to introduce a new designation, different from that which is presently used and which originates from Liebig and Poggendorff[?].

On the next day the congress met for the last time. Dumas opened the session with a speech in which he again tried to place an abyss between the old and the new, to settle artificially the business about designations, proposing to retain the old designation in inorganic chemistry and to accept the new weights in organic chemistry. In his opinion, the impossibility of applying the new ideas to mineral[16] compounds serves as the foundation for this proposal. Then Dumas characterized very well both existing directions. One, he said, represents a clear following of Lavoisier, Dalton, and Berzelius. The initial point for scholars of this school of thought is the *atom*, an indivisible, simple substance; all the rest is the sum of atoms, the magnitude of which is derived from the first atom. The other group goes along the path of Ampère and Gerhardt; it takes prepared substances and compares them; it takes molecules of a substance, discovers their transformations, and compares their physical properties. The first group did everything for mineral[16] chemistry; it is ineffectual in organic chemistry because here chemistry can create little from the elements. The second group, undoubtedly strongly promoting organic chemistry, has not done anything for mineral[16] chemistry. Dumas said, "We abandon both to proceed by our own paths; they must all converge." And in order to achieve agreement in the designations Dumas proposed accepting the new weights for organic substances and retaining the old weights for mineral[16] chemistry.

Cannizzaro and several other chemists took exception to Dumas' proposal. Cannizzaro's objection consisted of the fact that, having accepted the new ideas about molecules, we are not able to retain the old, Berzelius conclusions concerning weights and that if we arrive at those same weights by other paths, then we must clearly show that we are already guided by the new ideas. The new weights are not such news in the practice of science as to evoke strong opposition; everyone working in the new direction in England, France, Russia, Germany, and Italy, everyone, more or less, has already made use of Gerhardt's weights since they are based on a constant, involuntary principle. If we only correct some of Gerhardt's errors, we will obtain a consistent designation. Now these corrections of the Gerhardt weights are news at the present time, and it is impossible to demand their general introduction; but one must not lose sight of them.

The voting of the congress gave a positive answer to the desire for introducing changes in the weights.

The session was closed with the hope that in future years chemists will meet several times and will enter into the solution of the questions raised.

To this account I add the observation that in all the discussions there was not one hostile word between both parties. It seems to me that all this is a full guaranty for a rapid success of the new principles in the future. Of one hundred fifty chemists not one decided to vote against these principles.

Heidelberg, September 7, 1860 [D. Mendeleev]

Footnotes to Appendix IV

1 Sankt-Petersburgskie vedomosti (Saint Petersburg News), No. 238, 1860. — A. [For an official account of the sessions of the Karlsruhe Convention see Compte rendu des séances du Congrès international des chimistes réuni à Carsruhe le 3, 4 et 5 september 1860 in Mary Jo Nye: The

Question of the Atom, Tomash Publishers, Los Angeles, San Francisco, 1984, pp. 633–650 (English translation, pp. 5–28). — T.]

[2] Justus von Liebig (1803–1873). Professor of Chemistry at the Universität Giessen (1825–1851) and the Universität München (1851–1873). See F. L. Holmes: Justus von Liebig. In: DSB, 1973, Vol. 8, pp. 329–350. — T.

[3] Johann Christian Poggendorff (1796–1877). Professor Extraordinarius of Physics at the Universität Berlin (1834–1875), Editor of Annalen der Physik und Chemie (colloquially called Poggendorff's Annalen (1824–1877) and Biographisch-Literarisches Handwörterbuch zur Geschichte der exacten Wissenschaften, 2 vols., J. A. Barth, Leipzig, 1863, an indispensible biographical-bibliographical reference work known simply as Poggendorff. See Friedrich Klemm: Johann Christian Poggendorff. In: DSB, 1975, Vol. 11, pp. 49–51. — T.

[4] For a recent biography of this now relatively neglected scientist see F. W. J. McCosh: Boussingault: Chemist and Agriculturist, D. Reidel Publishing Company, Dordrecht, Boston, Lancaster, 1984. Also see Richard P. Aulie: Jean Baptiste Dieudonné Boussingault. In: DSB, 1970, Vol. 2, pp. 356–357. — T.

[5] Adolph Friedrich Ludwig Strecker (1822–1871), Professor of Chemistry at the Universities of Christiania (Oslo) (1851–1860), Tübingen (1860–1870), and Würzburg (1871). — T.

[6] Henry Enfield Roscoe (1833–1915). Professor of Chemistry at Owens College, now the University of Manchester (1857–1885). See Robert H. Kargon: Henry Enfield Roscoe. In: DSB, 1975, Vol. 11, pp. 536–539. — T.

[7] See footnote 6, Chap. 4. — T.

[8] The modern formulas for these molecules are H_2, H_2O, and H_2SO_4, respectively. — T.

[9] The modern formulas are H_2SO_4 and HNO_3, respectively. — T.

[10] The modern formulas are HNO_3, H_3PO_4, $H_2C_2O_4$, and KOH, respectively. — T.

[11] During the first half of the nineteenth century, measurement of vapor density (the Dumas method) was the only method available for determining molecular weights, and it was limited to gases or substances that could be vaporized without decomposition. Until the classic work of Raoult and van't Hoff on the colligative properties of solutions (1882 and thereafter) no reliable method existed for determining molecular weights of nonvolatile substances. — T.

[12] Hermann Kopp (1817–1892). Privat-Dozent (1841–1843), Professor Extraordinarius (1843–1852), and Professor of Chemistry (1852–1863) at the Universität Giessen and Professor of Chemistry at the Universität Heidelberg (1863–1890). See Henry M. Leicester: Hermann Kopp. In: DSB, 1973, Vol. 7, pp. 463–464. — T.

[13] Methane (CH_4) with ethylene (C_2H_4). — T.

[14] Mercury(I) chloride or mercurous chloride (Hg_2Cl_2) with mercury(II) chloride or mercuric chloride ($HgCl_2$). — T.

[15] Otto Linné Erdmann (1804–1869). Professor Extraordinarius (1827–1830) and Professor (1830–1869) of Technical Chemistry at the Universität Leipzig, he edited the Journal für praktische Chemie from 1834 to 1869. See W. H. Brock: Otto Linné Erdmann. In: DSB, 1971, Vol. 4, pp. 394–395. — T.

[16] Inorganic. — T.

Appendix V

An Extract from the Proceedings of the Conference of the St. Petersburg Medical-Surgical Academy of March 17, 1861 Concerning the Extension of Mr. Borodin's Stay Abroad to August, 1862[1]

The letter, addressed to the President, from Doctor Borodin, who is abroad for improvement, and his petition to the conference give an account of his future plans abroad. With the conclusion of the chemical congress in Karlsruhe [he] left Germany in order to spend the winter in Paris for the study of physics, physiology, and natural sciences during the current school year at the Collège de France [where] special courses are read, namely sections on physics and physiology which are most important for him. Regnault reads on caloric, a field which he enriched by numerous discoveries which are especially important for chemistry, Claude Bernard, on the physiology and pathology of blood. Thanks to the obliging Parisian chemist Riche,[2] who furnished him with the instruments, Borodin is studying polarization; moreover, he works at the Musée des Sciences Naturelles [(Museum of Natural

Sciences)], using its collections; he listens to various lectures in order to become familiar with the teaching methods of different professors. He attends meetings of the Paris Society, of which he was made a member. And since the plan conceived in [St.] Petersburg appeared to be impractical, he was obliged to give up a great deal and to constrain significantly the scope of his work. Whenever he was obliged to deal with the practical side of some subject, he was met with a great number of obstacles which required much time to overcome. Since the goal of his trip is practical instruction, he must learn a great number of things not having any relation to his purely chemical knowledge; he must learn a great number of analytical and other methods, which are necessary for him as a practical teacher. He faces the entire field of applied chemistry, physiological [chemistry,] and pathological chemistry. All this requires much time and a prolonged stay abroad in various educational centers of Europe; therefore [he] requests an extension of his stay abroad at least until the summer of 1862, especially since his activity in [St.] Petersburg at this time is less useful since the laboratory will not be ready yet and he is not able to begin a theoretical course because he must return in the middle of the school year. In addition, Mr. Borodin has presented three brochures on works published abroad in the Zeitschrift für Chemie und Pharmazie . . . [*B* 5–7].

The conference of the Academy, taking into consideration Mr. Borodin's scholarly works on organic chemistry and being in complete agreement with his opinion about the necessity of his further stay abroad for improvement in the practical aspects of specialties selected by him, decided to request the President for an extension until August, 1862 for Mr. Borodin with the same salary.

Footnotes to Appendix V

[1] Military-Historical Archives of the USSR in Leningrad, Stock No. 749, Military-Medical Academy, File No. 196 from 1861, Sheets 86–87. — A.

[2] Jean Baptiste Léopold Alfred Riche (1829–1908), Professor of Inorganic Chemistry at the École Supérieure de Pharmacie, Paris and assayer at the Mint. He was awarded the Gold Medal of the Académie de Médecine in 1880. — T.

Appendix VI

Borodin's Report of His Trip Abroad[1]

During all the time of my stay abroad I never lost sight of the fact that the Academy, having supplied me with the means for my final education in the sciences, had the goal of preparing me as a teacher of chemistry, which would meet the requirements of a contemporary education for physicians. Experiencing fully the entire gravity of moral debt resting upon me, I did my best in order to justify the trust of the Academy and to become worthy of the high calling – that of an instructor of youth in the field of science. Filled with the conviction that only a scientist who is completely possessed by this subject can be a really good teacher, I tried above all to develop myself from this point of view. This is accomplished first: by mastering what was done by others, and second: by independent research, helping the advancement of science. Without these two conditions it is impossible to obtain an accurate, critical outlook in science and to stand at the level of contemporary direction. But this is still not enough for the activity of a teacher: it is necessary to be able to teach others; it is necessary to be able to transmit science to audiences, conforming to their degree of development and to their future purpose. This is achieved, on the one hand, by the study of different methods of teaching others, and, on the other hand, by independent training. From the present report the Academy can ascertain only how I have used my time abroad and how much I have grown in science; successes in the profession of teaching can only be shown later on because the activity of a teacher requires a certain degree of experience and tact, which is acquired only with time by independent instruction. Abroad I first of all wanted to study those practical methods of chemical research which I did not have the opportunity of studying in Russia. For this goal I was sent to Heidelberg in order to study gasometric analysis and several other analytical methods in Bunsen's laboratory. But at this time all the places in his laboratory were occupied, and in order not to waste time I entered the laboratory of Privat-Dozent Erlenmeyer, who also had a section and equipment for gasometry. Besides, his laboratory had other advantages for me – for instance, with Bunsen I am only able to work until five, on Saturdays until three, and not at all on holidays and vacation time; with Erlenmeyer I can work as I please. The equipment in Bunsen's laboratory is set up exclusively for analytical work for beginners and for the study of analytical methods developed by Bunsen himself; with Erlenmeyer one can study many different things and can carry out serious scientific investigations. Finally, Erlenmeyer's personality itself if of more interest to me. Being the editor of the journal Kritische Zeitschrift für Chemie,[2] Erlenmeyer always continued on the level of contemporary direction in science and was occupied with organic chemistry, while Bunsen, limiting himself to a narrow frame of development of a few analytical methods for physico-chemical research, lost any interest in chemistry as a science (especially organic chemistry) and long ago fell behind in it. Consequently, I am able to make use of my time with far greater profit in Erlenmeyer's laboratory than in Bunsen's, although the latter benefits from great public authority. Later on I was satisfied in visiting Bunsen's laboratory unofficially and in observing the pursuits of some of my acquaintances who were working there.

In addition to the study of gasometric methods I became occupied with the study of methods of working with sealed tubes. With this goal in mind I undertook special work: the study of the action of zinc ethyl on chloroacetic and bromobenzoic esters. The idea of this work was: to try to find an efficient method for obtaining a complete series of new acids, isomeric with one another, by substitution of chlorine or bromine in chlorinated or brominated acids by hydrocarbon radicals. In connection with this work, making use of the low price of chemicals abroad, I began to look for a convenient method of brominating the fatty acids, butyric and valeric [acids]. In addition, I tried out many other new reactions.

I did not audit courses because all the university courses were very elementary, and the special courses which were of interest to me were not read at that time. However, I sometimes visited the lectures of Bunsen, Erlenmeyer, Helmholtz, and Kirchhoff, not so much to learn the subject as to become acquainted with the manner of instruction of these scholars; I also visited the lectures in order to see the performance of some of the more interesting experiments which I had not previously seen. While unofficially visiting Bunsen and Kirchhoff's laboratory, I observed their work which resulted in the discovery of spectral analysis and two new elements, cesium and rubidium.[3]

At the end of the winter semester I stopped my work in the laboratory, and in the spring of 1860 I left to inspect various laboratories [and] chemical and mining factories in Berlin, Holland, and south Germany. Returning to Heidelberg in the summer, I again resumed work in Erlenmeyer's laboratory. By the way, I visited some practical courses: botany with Schmidt[4] and geognosy [(structural geology)] with Leonhard.[5] Having finished my research on benzidine, I went to Karlsruhe in September 1860 for the chemical congress, which had the goal of clearing up several ambiguities in the conception of the chemical molecule, atom, etc. and finally to confirm the supremacy of the unitary system in chemistry [See Appendix IV. — T.]. After the congress I heft Germany and spent the following winter of 1860–61 in Paris, studying not only pure chemistry but also other sciences closely related to it. Owing to a fortunate accident, special courses were read this semester, namely in those branches of physics, physiology, mineralogy, etc. which are most important for me as for chemists and medical students. So for example: Regnault read *On Caloric*, Claude Bernard read *Blood and Other Liquids of Organisms*, [and] Sénarmont[6] [read] *Physical Properties of Crystals.* Regnault's course [was] serious and designated for people who already know elementary physics and mathematics and was interesting in the highest degree. This scholar's lectures were not only demonstrative but also philosophical. They were especially remarkable in the depth and fullness of critical opinion, richness of independent and new ideas, and clarity of presentation. One can understand all this if one remembers that the greater half of this branch of physics was created by the works of the same Regnault. Claude Bernard's course, less special and designated chiefly for physicians, was remarkable in the richness of new facts and unusual clarity and simplicity of presentation. Claude Bernard possesses the rare ability to transmit science in such a manner that each lecture is engraved in the memory of the listener without any effort of the latter's part. Sénarmont's course, also not very special, was very useful to me, specifically his demonstrative side, especially the subject, *On Thermoconductivity*, *Electroconductivity*, *Optical Indices of Crystals, etc.*, the fruit of the personal research of this scholar.

Further, I unofficially visited Sainte-Claire Deville's and Pasteur's laboratories in the École Normale, which is a secluded [closed] institution. My goal in this case was to inspect the works of these scholars, of which the first (Sainte-Claire Deville) was working on the determination of the specific gravity of vapors at high temperatures and the melting of refractory metals, platinum and others. Pasteur's works were even more interesting: this scholar was occupied at that time with his excellent research on fermentation and with the solution to the problem of spontaneous generation of animals. In addition to the three courses mentioned, which I attended regularly, I sometimes visited a great number of lectures in chemistry and natural science at the Collège de France, École de Médecine, École de Pharmacie, Sorbonne, École Normale, Jardin des Plantes, Conservatoire des Arts et Métiers, etc., solely with the goal of becoming acquainted with the manner of teaching of the French professors, because nowhere do they read lectures with such clarity and elegance as in France.

All this filled up my time to such a degree that it was impossible for me to work anywhere in a laboratory and especially to do any independent chemical research. However, I did have practical work; for example, at home I worked with polarization applied to chemical research. In this case I was much obliged to Professor Riche, who supplied me with all the necessary equipment, which was so expensive that I would never be able to acquire it myself. The other practical work consisted of blowing glass apparatus – something necessary for a chemist, especially with us in Russia; for this goal I took lessons from a mechanic, du Bois, who taught at the École de Médecine, the Sorbonne, and other institutions. Finally, as a member, I visited the meetings of the Paris Chemical Society and at them read the works which I had carried out in Germany. In spring during vacation I went to Italy to become acquainted with places of volcanic phenomena, with the formation and mining of sulfur, boric acid, etc. There I gathered a collection of lava from Vesuvius and Monte Somma and other mineral formations which are not found in our academic museum. By the summer of 1861 I went to Germany and stayed in Würzburg, having in mind to study physiological chemistry with Scherer, but after staying there 2 weeks I was convinced once and for all of the uselessness of my remaining in Würzburg and went to Giessen. The purpose of the trip to Giessen was to take practical crystallography with Kopp. From Giessen I made a small excursion for a survey of mines, chemical factories, laboratories, etc. in the neighboring towns of Germany. At the end of the summer I again returned to Heidelberg where I undertook work with the goal of studying the action of chloroiodoform on zinc ethyl [*B 12*]. In the middle of September I left for the Congress of Chemists and Naturalists in Speyer. During vacation in October I again went to Italy and this time exclusively for Cannizzaro, whose ideas produced an immense reform in chemistry by the development of the molecular theory and by the establishment of a precise conception of the weight of a chemical molecule. A quite unforeseen circumstance – namely a delay by the Prussian post of my fiscal money which was sent to me – caused me to remain in Pisa much longer than I estimated. In order not to waste time I began to work in the university laboratory. Being prejudiced, like all of us, against Italian universities, I planned to remain in Pisa only until I received my money and therefore only to carry out a few analytical works which could be stopped at any time. But from the first day I saw that the Pisa laboratory offered immeasurable advantages over other laboratories. This laboratory is not public and therefore is not organized on those mercantile-commercial foundations as are the German laboratories. Piria[7] and Bertagnini,[8] having made almost all of their discoveries in this laboratory, had time to enrich it with many pieces of equipment and other scholarly instruments. Italian scholars have not yet become accustomed to an influx of foreigners who arrived for the purpose of working, and, with quite different experiences than the German scholars, they have not become accustomed to that system of exploitation which is leading science to the level of a craft. Therefore it was natural that the professors of Pisa University, de Luca and Tassinari, not only accepted me with the highest degree of kindness but immediately offered me the use of all the laboratory facilities. Finding a rare stock of platinum vessels there, I immediately gave up the analytical work, and, taking advantage of such a fortunate opportunity, I undertook serious work with fluorine compounds, which I had never studied because of the lack of means. These compounds are extremely interesting and have not been studied to any degree. The reason for this is found in the special property of fluorine compounds to enter into reactions with almost every substance; as a result they attack glass and porcelain vessels; they also form double compounds at each step, which greatly impedes the purification and analysis of the fluorine products. Work of this kind must be carried out only in a platinum vessel, which is very expensive and not readily accessible. All this taken together explains why fluorine compounds were studied so little until now in spite of the fact that their study presents an immense interest to science. The absence of any kind of entertainment in Pisa, the availability of the means for study, the rich museums and collections, the libraries, and finally, the inexpensive living conditions and good climate all are favorable in the highest degree to studying, and I dare say that I accomplished more for science and for education in Pisa than where I would have been. Much time was left for me to pursue other sciences, especially physics with the well-known scholar, Felici.[9] In addition to work with fluorine compounds, I also carried out two original studies with benzil [*B 11*] and chloroiodoform [*B 12*]. Having published all of this in the May issue of Il Nuovo Cimento, I said goodbye to Italy and proceeded north, to Germany. After this I no longer worked at anything seriously, I did not take any courses and looked at only what I did not have time for previously. I collected various references and information, having in mind the acquisition of equipment and materials for the Academy laboratory. I purchased many things myself, especially the precision instruments which one cannot order but which must be selected and examined in person.

From the present account the Conference can see that I studied primarily pure organic chemistry and that I did not work at all in those so-called chemico-physiological or chemico-medical laboratories. Perhaps many would reproach me for this, thinking that I did not want to work in such laboratories because of a lack of interest in chemico-physiological or chemico-medical research, which in the opinion of many is much more difficult work than pure chemistry. In order to prevent such an inference and, at the same time, to indicate the real reason for this circumstance, I must say that at first I really intended to work in such laboratories, but later I saw that it was useless. All of these laboratories could be divided into two categories: one category is established primarily for medical students who have not received a serious education in chemistry and who have not become at all familiar with the technical side of chemical work. The equipment of such laboratories is intended for elementary analytical work, especially for the analysis of urine by means of titration. All the analytical methods offered there, and presenting serious difficulty for medical students who have never studied chemistry, do not present anything new and difficult for the chemist who has been brought up on pure chemistry. Finally, the chemist has not been satisfied for some time by the degree of purity, accuracy, and precision which the works of these laboratories produce.

The second category consists of those so-called scientific chemico-physiological laboratories. Their goal is the use of chemistry for the solution of physiological problems. The problem in this case is more difficult and more complex, but the chemist meets serious difficulties here only in the physiological part of the work; the chemical side consists for the main part of those very general analytical methods which were already well known to the chemist long ago. In order to overcome the difficulty presented by the physiological side of the work it is necessary to devote much time to physiology, to abandon pure chemistry completely, and to become a physiologist. I did not have the privilege or inclination to deviate to such a degree from my special assignment . . .

From the papers published abroad by me, I refer first to those which constitute the fruit of independent laboratory work. They are:

1. Über die Wirkung des Jodaethyls auf Benzoylanilid (published in Zeitschrift für Chemie und Pharmazie, August, 1860) [*B 3*].
2. Über die Amide des Benzidins (in the September issue of the above journal) [*B 6*].
3. Über die Monobromvaleriansäure und Monobrombuttersäure (in the December issue of the above journal) [*B 7*].
4. Über die Wirkung des Zinkaethyls auf zusammengesetzte Aether (ibid.) [*B 8*].
5. Fatti per servire alla storia di fluorure (in the May issue of the journal Il Nuovo Cimento, 1862) [*B 10*].
6. Fatti per servire alla storia di benzile (in the June issue of the above journal) [*B 11*].
7. Sull'azione dello zincoetile sur chloroiodoforme (ibid.) [*B 12*].

In addition to these, I published several articles based on my discoveries of the compounds described in previous papers. These articles are:

1. Recherches sur les acides bromes (read by me in a meeting of the Paris Chemical Society and printed in one of the booklets of the Bulletin of the society) [*B 7*].
2. Beiträge zur Geschichte des Fluorurs [*B 10*].
3. Beiträge zur Geschichte des Benzils [*B 9*].
4. Beiträge zur Geschichte des Chloroiodoforms (all three articles are published in the journal Zeitschrift für Chemie und Pharmazie, 1862) [*B 12*].

5. Sur le fluorure de benzoyle (in Comptes rendus hebdomadaires des séances de l'Académie des Sciences à Paris, 1862) [*B 10*].

For lack of separate reprints of all of these articles I am able to attach to the present report only an extract from the journal Il Nuovo Cimento.[10]

31/I 1863 [January 31, 1863] A. Borodin

Footnotes to Appendix VI

[1] Military-Historical Archives of the USSR in Leningrad, Stock No. 749, Military-Medical Academy, File No. 12 from 1858–1863, Sheets 103–116. — A.

[2] Erlenmeyer edited this journal (with G. Lewinstein) from 1859 to 1864. It ceased publication in 1871. — T.

[3] See footnote 3. Appendix III. — T.

[4] Johann Anton Schmidt (1823–1905) studied botany at Heidelberg (1848–1849) and Göttingen (1849–1850, PhD 1850). At the Universität Heidelberg he was a Privat-Dozent (1852–1855) and Professor Extraordinarius of Botany (1855–1863). — T.

[5] Karl Cäsar von Leonhard (1779–1862), Professor of Mineralogy at the Universität Heidelberg (1818–1862). See John G. Burke: Karl Cäsar von Leonhard. In: DSB, 1973, Vol. 8, pp. 245–246. — T.

[6] Henri Hureau de Sénarmont (1808–1862), Professor of Physics at the École Polytechnique (1856–1862). See Walter Fischer: Henri Hureau de Sénarmont. In: DSB, 1975, Vol. 12, pp. 303–306. — T.

[7] Rafaelle Piria (1815–1865), one of Dumas' students, was Professor of Chemistry at the University of Turin. — T.

[8] Cesare Pietro T. Bertagnini (1827–1857), Professor of Chemistry at the University of Pisa (1856–1857), had worked with Liebig and visited London, Paris, and New York. — T.

[9] Riccardo Felici (1819–1902), Professor Extraordinarius (1856–1859) and Professor (from 1859) of Physics at the University of Pisa. — T.

[10] For a translation of the names [and for complete reference citations] of Borodin's articles see Borodin's Chemical Works. — T.

Appendix VII

On the Appointment of Assistant Professor Borodin as Full Professor

At the meeting of the Conference of the Imperial Medical-Surgical Academy on the past March 28, on the occasion of the expiration of the 30-year period of service of Academician and Honored Professor Zinin's teaching position, it is resolved: upon his release from the stated position to ask for a petition to confer upon him the duties of director of the practical work in chemistry; whereupon the Conference, having in view the fact that the teaching of theoretical chemistry with preparation for lecture demonstrations, with the recent growth of science, and with the close bond with medicine, requires much more work than lecture time alone from the teacher, determined: after Zinin's release to attend to the replacement of the vacant Chair of Chemistry on the basis of § 36 of the Regulations of the Academy.

On April 3 by Imperial assent, the Honored Professor Zinin was retained for service at the Academy in the future at the rank of director of chemical works with the rights and salary of full professor.

Assistant Professor Borodin was suggested by me and by the Honored Professor Zinin as the candidate for the vacant Chair of Chemistry.

Doctor of Medicine Borodin, having completed the course of medical science at the Academy in 1856, was awarded the degree of physician with honors and was rewarded by the list of commendation; on March 25, 1856 he was appointed a supernumerary intern in the 2nd St. Petersburg Military Land Forces Hospital and was assigned as assistant at the diagnostic clinic of Professors

Zdekauer and Besser. During this time he also worked in the chemical laboratory of the Academy; on June 26, 1857 he was sent abroad for four months with the commission to inspect foreign chemical laboratories and to acquire diverse chemical apparatus for the Academy's laboratory; upon his return from abroad he was attached to the Department of Chemistry with the rank of assistant, and during the academic year 1857–58 he directed students of the 2nd course in practical studies of chemistry in the Academy's laboratory; on May 3, 1858 he publicly defended the dissertation: Ob analogii mysh'iakovoĭ kisloty s fosfornoiu, v khimicheskom i toksikologicheskom otnoshenii (On the Analogy between Arsenic Acid and Phosphoric Acid in Chemical and Toxicological Behavior) [*B 2*] and was awarded the degree of Doctor of Medicine; during the vacation of 1858, owing to an offer of the members of the Military-Medical Academic Committee, he left for Soligalich for a chemical and medical investigation of the mineral waters there. These investigations were published in various periodicals and then in a separate brochure entitled: Opisanie Soligalicheskikh mineral'nykh vod (Description of the Mineral Waters of Soligalich) [*B 4*]; during 1858–59 he directed the practical chemical studies of young physicians who remained at the Medical Institute of the Academy for improvement; at the same time he read the course in chemistry applied to physiology and pathology and the course in the history of the development of chemical theories to young physicians; on October 6, 1859 he was sent abroad by the Academy for improvement in chemistry, at first for 2 years, but then he remained abroad until August 1, 1862; while abroad he studied in the chemical laboratories of various universities, and he carried out several original works, which were published in various special foreign journals; in 1860 he was elected a member of the Paris Chemical Society; on September 13, 1862, upon his return to Russia, he was appointed to fill the duties of assistant professor of chemistry, and on December 8 of the same year he was confirmed as assistant professor; during the 1862–63 academic year he read the complete course in chemistry at the Academy for the 1st course in inorganic and the 2nd course in organic chemistry. In addition to this, in the fall of 1863 he was invited and confirmed as professor of chemistry at the Forestry Academy, where he also read the complete course in chemistry.

During this period of time, in addition to his toxicological work with phosphoric and arsenic acids, which served as the theme for his doctoral dissertation, and his investigation of the mineral waters of Soligalich, he carried out many independent, original works in chemistry . . .[1].

No other candidates have been proposed to occupy the vacant chair of chemistry.

A secret ballot was taken at the meeting of April 11th in accordance with the regulations which had imperial approval on July 26, 1861; Borodin received 17 positive votes and one negative vote.

The Conference of the Academy, having in view the fact that Borodin received a majority of positive votes and that he fulfilled all the conditions required by the regulations of the Academy for receiving the rank of full professor, decided: to appoint Doctor Borodin full professor of chemistry with a salary appropriate for this position.

President P. Dubovitskiĭ

Footnote to Appendix VII

[1] A list of Borodin's works published during the period 1858–1862, follows. — A. [See Complete List of Borodin's Chemical Works. — T.]

Appendix VIII

Borodin's Speech Delivered on February 9, 1880 at N. N. Zinin's Funeral[1]

I do not direct my speech to the body of the deceased teacher; it is deaf and dumb . . . I direct my words to you who have gathered here to honor the memory of the deceased. You read the words: Grandfather of Russian Chemistry, on the garland which the students carried. But do you all know how great are the merits which secure this honorable title for the deceased? We will recollect these merits before we cast a handful of earth on the good teacher's coffin . . .

Possessing vast knowledge, a profound, lucid mind, and ardently loving Russia – Nikolaĭ Nikolaevich understood earlier than others that we will not have science at home, young energy will not be attracted to cultivate it, here in the heart of the motherland, as long as no independent Russian school has been started. And here in Kazan he, for the first time, laid the foundation of the Russian school of chemistry. The seed, scattered on grateful soil, took hold and sprouted great roots. The youthful Kazan school developed quickly, grew, became stronger, and spread its shoots far out to other breeding grounds of knowledge in Russia. A pupil of Nikolaĭ Nikolaevich became, in turn, a teacher and continued his teacher's work.

It is not the fate of every public figure to see the fruits of his activities, the complete development of the work established by him. In this respect an enviable fate fell to Nikolaĭ Nikolaevich; the scholarly activity of three generations of the school that be had created was conceived, developed, and flowered before his eyes; he saw not only his children of science but [also his] grandchildren, and even great-grandchildren, who are now those youthful scholars of the large family of Russian chemists to whom the future of science belongs. The moral satisfaction which the deceased must have experienced from this was all the more complete as he lived to an old age but did not outlive himself as a scientist. He did not stand still at any period of his activity nor did he grow cold during any phase of the development of science. Despite the extremely rapid growth of chemistry he did not fall behind its contemporary movement. Strong of body and soul, until the past year he continued to work tirelessly in the laboratory and to enrich science by new discoveries. Until the last days when he was ill and confined to his bed, he did not stop reading, following with vivid interest the successes of science, understanding and taking to heart all that was done by the young generation of chemists.

The latest period of his scholarly activity does not belong to Kazan but to [St.] Petersburg, where he was invited to a chair in the Imperial Medical-Surgical Academy. Since I had the great honor to be the personal apprentice of the deceased, then the junior colleague, and, finally, the successor to [his] chair, I consider it my right, which is dear to me, to recall here his academic activities. Entering the staff of professors of the Academy, he brought there those same resolute and high principles of science, progress, and independent action. His words from the rostrum were not only a reliable transmission of the contemporary state of science but also a tribune of a new direction. As a worker in the sphere of medicine, he steadfastly supported the idea that medicine, as a science, offers only an application of natural science to the preservation and restoration of the health of man, and therefore the natural sciences must play the role of paramount, basic subjects in a medical education and not be supplementary or auxiliary; he also believed that the medical student should master not so much the fragmentary facts of applied natural science as the general system of science, the methods of thinking, and the methods of research of the naturalist; therefore the teaching of natural sciences for the medical courses must be well-grounded and as complete as possible, not limited to applied information; he believed that for a conscientious conception of how science is formed and for a clear assimilation and true appraisal of what is done by others in science, it is necessary to do some work in the field of science and to contribute one's part, even if small, to the general storehouse of knowledge.

The deceased adhered to these ideas in all of his various academic activities. Thanks to his energy and persistence, the teaching of chemistry at the Academy was placed on a solid foundation; a new, spacious chemical laboratory was constructed, which was supplied with all the educational equipment in accordance with contemporary requirements; the practical work in chemistry was arranged on a large scale for students, physicians, and pharmacists. Faithful to the ideas which he had already conceived in Kazan, he was not restricted by the confined limits of the practical work of chemistry, but he tried to achieve even higher goals: he endeavored to foster a serious love for science, to challenge Russian youth to independence, and to found here a new breeding ground for the Russian chemical school, naturally, as far as the special character of the Academy allowed. The conditions at this time promoted the realization of Nikolaĭ Nikolaevich's cherished wishes. It coincided with one of the most lucid epochs of academic life: with the time of large-scale reforms. Thanks to a fortunate coincidence, people who were inspired by similar ideas and who developed them in those reforms which placed the Academy at its proper stature, were at the head of the Academy's administration. These reforms consisted: (1) of the attraction of fresh teaching strength; (2) of the organization of academic-auxiliary institutions for the development of practical teaching of various fields of medicine and natural science as extensively as possible; and (3) of the establishment of the "Institute of Young Physicians" as a breeding ground for the growth of scientific strength in the field of Russian medicine. An ardent patriot, Nikolaĭ Nikolaevich once again experienced the high moral satisfaction of seeing also here [in St. Petersburg] the fruition of those ideas which he had cherished for the Russian man [and] which the deceased had placed at the foundation of his many years of scholarly activity. The Institute of Young Physicians, established by persons placed at the head of the Administration of the Academy at the end of the fifties, accomplished on its own the achievement of becoming a breeding ground of Russian scholars and educational strength not only in the field of medicine but partly also [in] natural science. In the course of 20 years it gave Russia 129 young public figures, of whom 58 devoted themselves to professorial activity at academies and universities; the rest distinguished themselves more or less prominently by activities in the field of medical science, either practical or administrative.

After what was expressed by my esteemed comrade A. M. Butlerov, need I say more about Nikolaĭ Nikolaevich as a man? Is it necessary to add that the deceased made use of this seed of goodness, truth, and knowledge with infinite respect and love?

The students surrounding him became accustomed to seeing him as an ardent defender of their serious interests, an experienced adviser and leader, a reliable assistant, and a friend.

His infinite kindness, accessibility, affability, simplicity, and warmth in the treatment of people and his readiness and ability to help anyone who needed it made the famous name of Zinin one of the most popular in the Academy.

His memory is inseparably linked with the very lucid recollections of academic life.

The loss of this man is felt all the more.

Footnote to Appendix VIII

[1] [From the] journal Zdorov'e (Health), 1880, No. 135. — A.

Appendix IX

Borodin's Letter of March 13, 1883 to P. P. Alekseev[1]

Sincerely and deeply respected Pëtr Petrovich, I venture to send you the subscription list for the collection of donations for the memorial to N. N. Zinin. I turn to you for this with a humble request, as an old comrade and representative of chemistry in Kiev: to assist in the dissemination of signatures in the university circle and any other place you find necessary. Generously forgive my unceremonious appeal to you. The fervent wish that the Russian people honor the memory of one of the most renowned champions of the independence of Russian science and thought, our great chemist, serves as my excuse. It would be a shame if the Russian people, who supported the memorials of Liebig, Wöhler, and Claude Bernard, would remain deaf to the appeal to do the same for the deceased N. N. Zinin . . .

Borodin's Letter of March 13, 1883 to N. N. Beketov

Sincerely and deeply respected Nikolaĭ Nikolaevich, I venture to send you the subscription list for the memorial to the deceased N. N. Zinin. Knowing your regard toward the deceased and his regard toward you, I dare to hope that you will not refuse my request to assist in the dissemination of subscriptions in the university circle, the Veterinary Institute, etc. Russian society was invited to subscribe to the memorials of foreign scholars — Liebig, Claude Bernard, and recently Wöhler. It would be a pity if it did not answer the invitation to honor the memory of our celebrated scholar.

Borodin's Letter of March 14, 1883 to A. M. Zaĭtsev

Sincerely and deeply respected Aleksandr Mikhaĭlovich, I venture to send you the subscription list, and I do not doubt that the thought of honoring the memory of the venerable "Grandfather of Russian Chemistry" is sympathetic to you. Therefore I turn to you, the deserving successor of the chair on which the dear, departed one formerly won fame, with the great request to render assistance in the dissemination of subscriptions in the university circle.

Footnote to Appendix IX

[1] From materials collected by S. A. Dianin for Volume IV of Borodin's letters. — A.

Appendix X

Report to the Head of the Medical-Surgical Academy[1]

In view of the approaching date of the twenty-fifth anniversary of my tenure in government service, I have the honor to ask your Excellency to petition for inclusion in my pensionary period for educational service the entire time of my stay in service before the approval for regular appointment in the Academy.

As the basis for my request I have the honor to cite that circumstance that I, even from my first entrance into service, carried out, according to the appointment of the Academy, the teaching duties, guiding the studies of the students of the Academy and of the young physicians and reading to the latter a course in chemistry, etc., and I also carried out, according to the charge of the Academy, the duties related to the educational and scientific services which are usually charged to the regular teachers. These are:

1. From the file: O prikomandirovanii lekaria Borodina k professoru Zdekaueru, dlia ispravleniia pri nem obiazannosti assistenta (On the Attachment of Physician Borodin to Professor Zdekauer for Execution of the Duties of Assistant) (1856, No. 618) it is evident that after finishing the course I was appointed to the service as intern in the 2nd Military-Land Forces Hospital and was attached to the Academy with the responsibility for supervising the technical exercises of the students in medical diagnostics, in the capacity of assistant.
2. From the file: O sluzhbe ad'iunkt-professora Akademii Borodina (On the Service of Assistant Professor of the Academy Borodin) (1862, No. 155) it is evident that I also worked in the chemical laboratory of the Academy; on June 26, 1857 I was sent abroad for 4 months with a commission to inspect foreign chemical laboratories and to acquire various [items of] chemical apparatus for the laboratory of the Academy; on my return I was attached to the Department of Chemistry of the Academy at the rank of assistant and during 1857–58 I directed students of the 2nd course in practical chemical work in the laboratory of the Academy; owing to a proposal of the members of the Military-Medical Scientific Committee, I was sent to the town of Soligalich during vacation time for an investigation of the mineral waters there. In the next, 1858–59 [academic] year I directed the practical chemical work of young physicians who remained at the Medical Institute of the Academy for advanced courses; at the same time I read the course in chemistry applied to physiology and pathology and the history of the development of chemical theories to the young physicians. On Oct. 6, 1859 I was sent abroad for improvement, and on my return to Russia I was appointed to fulfill the duties of assistant professor of chemistry.
3. On February 5, 1872 I carried out the duties of academic secretary (see the repo[rt] of the conf[erence], 1872. — refer to the documents for the date).
4. On September 5, 1872 I was confirmed by the Minister of War as a teacher of chemistry for the Women's Courses for Training Midwives at the Med[ical]-Sur[gical] Acad[emy] with a salary of 700 rubles a year (see the docum[ent] in my name of October 11, 1872, No. 3027).
5. In accordance with the decision of the Conference of April 28, 1879 I was nominated to attend the Congress of Naturalists and Physicians in Kazan in the capacity of deputy from the Academy (see file No. 155–1862).
6. On June 26, 1873 I received written thanks from the Minister of War for successful teaching of the Women's Courses (see the docum[ent] in my name of June 26, 1873, No. 1882).
7. On January 11, 1876 I was confirmed by the Minister of War for the following three years as a member of the Economic Committee of the Academy (see file No. 155–1862).
8. On April 21, 1878, by a nomination of the Minister of War, because of the illness of the Chief of the Academy I fulfilled the duties of the latter until his recovery (N. B.: It is necessary to check the time period). (Note: see the docum[ent] in my name of April 21, 1878, No. 792).
9. On February 13, 1879 I carried out the duties of academic secretary (until Prof. Dobroslavin's return from a mission) (see docum[ent] of February 13, 1879, No. 217).
10. On April 19, 1879, by nomination of the Minister of War, I became a member of a committee to review the design of the regulations and the staff of the Military-Medical Academy (see doc[ument] of April 19, 1879, No. 750).

11. On September 25, 1879 I was confirmed as a candidate for membership of the Academic Court for 1879–80 (doc[ument] No. 3189).
12. I was confirmed as a member of the Academic Court for the current year (check documents). In addition, I have the honor of attaching two of the cited files and documents as well as the official list which was presented to me.

March 26, 1881 Academician A. Borodin

Footnote to Appendix X

[1] Branch of the Central State Military-Historical Archives in Leningrad, Stock No. 749, Military-Medical Academy, File 155 for 1887, Sheets 134–135. — A.

Appendix XI

Testimonial of Borodin Concerning the Works of D. I. Mendeleev and A. G. Stoletov[1]

The committee, having examined the scholarly works of the two candidates who were proposed for the vacant chair of physics – Professor of [St.] Petersburg University D. I. Mendeleev and Professor of Moscow University A[leksandr] G[rigor'evich] Stoletov, has the honor of presenting its conclusion. The works of the two are quite different in respect to substance because they concern quite different fields of science; Professor D. I. Mendeleev's work belongs to the study of heat and that part of physics which is found in closer contact with chemistry; Professor Stoletov's work [belongs] to the field of electrostatics, electrodynamics, and magnetism. With respect to quality and significance for science, the works of both represent serious and independent research, enriching science with new and important discoveries.

On this basis both candidates, in the opinion of the committee, are completely worthy of occupying the chair of physics, and each of them, in all fairness, can be called an adornment to our Academy.

September 16–17, 1873

Footnote to Appendix XI

[1] The original is preserved in the Archives of S. A. Dianin. — A.

Appendix XII

From Borodin's Report to the Conference of the Military-Medical Academy[1]

The present state of the Department of Chemistry in the Academy is a consequence of circumstances which have been established historically.

The fact is that the conditions of teaching chemistry in the Academy are substantially different from those in the universities. In the universities, in addition to the staff of the Medical Faculty, the staff of the Physical-Mathematical Faculty also takes part in the teaching of chemistry to medical students. Only one medical or physiological chemistry concerning practical work is given by the Medical Faculty. The rest of the fields of chemistry are given by the Physical-Mathematical Faculty where the medical students hear the general course in inorganic chemistry, organic [chemistry] and analytical chemistry together with naturalists, and only by way of an exception are separate parallel courses given to medical students.

So, for example, in Moscow, Kharkov, Kazan, and Warsaw, courses in general chemistry are given, and in Kiev Professor [Fedor Minich] Garnich-Garnitskiĭ [(1834–1892)][2] gave a special parallel course in inorganic chemistry to medical students. Thus the teaching is carried out by at least three professors who have at least one assistant; and in Kazan, for example, even four teachers read the various fields of chemistry to the medical students:

Inorganic chemistry – Prof. [G. N.] Glinskiĭ
Organic chemistry – Prof. [Aleksandr Mikhaĭlovich] Zaĭtsev [(1841–1910)]
Analytical chemistry – Prof. Flavitskiĭ[3]
Physiological chemistry – Prof. Shcherbakov

and each of them has at least one assistant.

We have always had abnormal conditions concerning the assistant staff in the Chemistry Department of the Academy. So we had two professors, Zinin and I, one assistant [professor] N. V. Sokolov and *one regular* assistant for all three of us – Zibert. After Zinin's departure two professors remained: N[ikolaĭ] V[asil'evich] Sokolov [(1841–1915)] and I and still only one regular assistant – P. G. Golubev.

Everyone knows how much work a laboratory assistant has in a Chemistry Department, especially one such as ours, where the practical work is conducted on a large scale. I suggest that you visualize the position of two professors, for example, of surgery or obstetrics, if they had only one assistant or orderly for both of them! And how the position of this assistant or orderly must be, acting as the helper of the two professors at the same time! This abnormal condition resulted in constant applications of the chemistry professors for an increase in staff even if only by one laboratory assistant. But for lack of means these applications always proved to be unsuccessful. In order to get out of the difficult situation nothing remains for the professor of chemistry but partly to carry out the laboratory duties himself (as for example, N. V. Sokolov and I for N. N. Zinin even now) and partly to place the duties upon people who are attached to the department by chance, or in any case, unofficially, who work voluntarily and free of charge out of kindness or out of an interest in science. However, what would be easy for a Physical-Mathematical Faculty where chemistry is a special subject is feasible only with difficulty in a medical establishment, as for example in our Academy, where students or physicians who are sufficiently prepared in chemistry for the most part pursue some medical specialty immediately. Only in very rare, exceptional cases does a student of medicine take so great an interest in chemistry that he acquires the training necessary for a laboratory assistant and desires to perform the duties of a laboratory assistant, a job which does not bestow any special advantages for a medical student in the future. I found such a laboratory assistant in the person of A. P. Dianin, who is one of my most gifted students and who, while still a student, studied chemistry so well that I was able to entrust to him the duties of laboratory assistant for the practical work of analytical chemistry. This was in 1873. From that time until now he has been acting as an assistant who is not on the regular staff in this field of teaching. For some time he carried out these duties free of charge; later on, thanks to the kind attention of the Head of the Academy, he received a small compensation from the residual amount [of money]. Recently it was decided, according to a resolution of the Conference, confirmed by the Minister, to make this compensation from the budget allocated to the positions. A. P. Dianin became an excellent, experienced, and talented teacher and laboratory assistant as a result of his continual work with us since 1873. Without such an assistant the practical work in analytical chemistry would be inconceivable for us where the number of workers is cumbersome and the personnel and budget of the laboratory is small. Compulsory hours of work from 5 to 7 [P. M.] are far from sufficient, and laboratory work continues for practically the entire day; many students have other studies during the hours prescribed by schedule, for example, anatomy, etc., and besides this, it is impossible to perform many of the complex chemical techniques in a *two-hour* period; finally, the course of chemical operations cannot be subordinated to such a rigid determination of time; on the contrary, one is obliged to consider the *working time* for each individual case, since, for example, it is impossible to collect some precipitate when it is not completely washed or to calcine it when it is not dry, etc.; it is impossible to determine precisely the times for these operations; consequently, one is obliged to perform everything at different times, whose limits must be broad enough and which practically consume the entire day. Moreover, workers cannot remain completely without supervision and instruction in order to avoid a futile waste of time, labor, materials, and vessels and to eliminate the possibility of explosion, poisoning, etc. from careless or ignorant use of dangerous chemicals, by means of which even the most elementary works are performed. Therefore the closest supervision and guidance in the practical work rests upon the assistants, of whom there are several in the majority of cases, but not one, when the number of workers is great, as for example with us in the Academy. Since one must introduce the practical work of beginners with some theoretical instructions and explanations, it is an extreme inconvenience to transmit the necessary information to each student individually when there is a large number of students. Hence the necessity for some type of special demonstrative reading or lecture results, which cannot agree with the theoretical course of the professor either in time, in system of presentation, and or in content. Everywhere such *practical* lectures always rest on the nearest and immediate intructors of the practical work, i.e., on the *assistants* of the department. By appointment of the conference such lectures were read by A. P. Dianin. The necessity for such readings was felt earlier; such was Assistant Professor Shalfeev's course: O metallakh (On Metals) at the time when the theoretical course of inorganic chemistry was given by N. V. Sokolov, but at that time the practical work in analytical chemistry was placed in the second course where the students were already acquainted with inorganic chemistry. As now prescribed, this work is transferred to the 1st course where the students do not yet know theoretical inorganic chemistry. Transfer of the practical work to the 1st course was done for two reasons: (1) an absolute lack of time of the students of the 2nd course since practical work in physiological chemistry, which formerly did not exist, was established; (2) the second reason, the necessity of preliminary preparation in *analytical chemistry* for work in *physiological chemistry*. The course in inorganic chemistry conforms to the demands of *theory*, the course in analytical chemistry to the demands of *practice*. The course in inorganic chemistry begins with metalloids and ends with metals, but the system of analytical chemistry [is] the reverse. Besides, the volume and range of information as well as the method of presentation are completely different in both fields of chemistry. Finally, with us, in view of the fact that mineralogy and crystallography are not read, the necessary information from the two mentioned sciences must be communicated during the presentation of the study of *metals for analytical chemistry*. Thus, if the need for a separate reading of: O metallakh (On Metals) by lecturer Shalfeev was felt in former times, this need is now inevitable. In the present case no one can satisfy this need better than A. P. Dianin, since he himself was our student and from 1873 he worked constantly with our students; he therefore knows all of the conditions, the way of life, and the needs of the students and all of the conditions and requirements of the Chemistry Department. In this respect A. P. Dianin is irreplaceable.

I add that besides students who are working on a large scale, we also have a great number of physicians who are doing their special work with other professors but are carrying out the chemical part of their work in the chemistry laboratory. These physicians require advice, help by work and deed, and [they] take up a great deal of the assistant in the Department of Chemistry's time. In this sense A. P. Dianin is an irreplaceable instructor, adviser, and assistant. He must often give more time to the work with the physicians than to the work with the regular students.

It is evident from all of this that A. P. Dianin must be retained *as assistant in the Department of Chemistry* until we get a permanent position for the second assistant and that, in the interests of the Academy, he should be entrusted with teaching on the same terms as previously established.

May 31, 1886 Academician A. Borodin

Footnotes to Appendix XII

[1] From material collected by S. A. Dianin for Vol. IV of Borodin's letters. — A.

[2] Known in the West as Theodor Harnitz-Harnitzky, in 1859 he obtained what he called chloracetene, shown in 1870 by Kekulé and Zincke to be a mixture of aldehyde, paraldehyde, and carbonyl chloride. — T.

[3] Flavian Mikhaĭlovich Flavitskiĭ (1848–1917) worked with Butlerov from 1870 to 1873 and taught at the University of Kazan from 1873 until his death. He was the first to interconvert monocyclic and bicyclic terpenes. See GSE, 1981, Vol. 27, p. 260. — T.

Appendix XIII

1. Borodin's Letter to D. I. Mendeleev

Mendeleev, old chap, one day you told me about a translation of Gerhardt and Chancel[1] with a supplement on the analysis of urine and other medical matters. If you are able to provide me with this work, I will be extremely grateful to you as I am in need of money at the present. Write [me] a couple of words. This chapter, you see, is not large, and one is able to manage it quickly. I'm giving you my address in case: Sampson'evskiĭ Avenue opposite the 2nd section of the Hospital in the Klimovs' house. My sincere regards to Feozva Nikitishna[2] and kiss your production with her.[3]

2. VI 1863 [June 2, 1863] A. Borodin

2. Borodin's Letter to D. I. Mendeleev

Thank you, old chap, for not forgetting me; I wasn't able to be with you yesterday in any case. Now I am writing to you to ask for the work which you have (even Gerhardt and Chancel) for a certain Svitenko, your former laboratory assistant, who is now living in poverty and is miserable. Perhaps I will assume the responsibility of the editorship if it is necessary. Svitenko will come to you today. Be in good health.

25. X 1863 [October 25, 1863] A. Borodin

Footnotes to Appendix XIII

[1] He refers to Analiticheskaia khimiia (Analytical Chemistry) by Gerhardt and Chancel, translated, supplemented, and published under the editorship of D. Mendeleev, Obshchestvennaia pol'za (Public Benefit), St. Petersburg, 1864–1866. — A. [The French book from which the translation was made was Charles Frédéric Gerhardt and Gustave Charles Bonaventure Chancel: Précis d'analyse chimique qualitative, 2nd edition, Victor Masson, Paris, 1862. — T.]

[2] Feozva Nikitishna Mendeleeva (née Leshchevaia), Mendeleev's first wife, whom he married in 1862. — T.

[3] This prebably refers to the Mendeleev's first child. They had a son and a daughter. — T.

Index of Borodin's Musical Compositions

Index of Names

Index of Subjects

E. Fischer

Aus meinem Leben

With a Prologue and an Epilogue by B. Witkop

Reprinted from E. Fischer, Gesammelte Werke
Edited by M. Bergmann

Reprint der Erstauflage Berlin 1922.
1987. (XXVIII) VI, 219 Seiten.
ISBN 3-540-18002-8

Emil Fischer was one of the most prolific researchers and teachers in chemistry in his period, i.e., the second half of the 19th and the beginning of the 20th century. In those days, it was still possible for him to encompass chemical sciences as a whole.

His studies in the fields of peptides and proteins, lipids and carbohydrates were epoch making, and established biochemistry as a discipline in its own right. His works are still valid today.

This book is a reprint of Fischer's autobiography edited by M. Bergmann, published in 1922 as part of Fischer's Collected Works. It gives an impression of Fischer's brilliance, his likeable personality, and the social life at the end of the 19th and the beginning of the 20th century. The Prologue and Epilogue in English by B. Witkop, Member of the National Academy of Sciences, Washington/DC, provide a link with today's modern science.

Springer-Verlag
Berlin Heidelberg New York
London Paris Tokyo